Advances in Oil and Gas Exploration & Production

Series Editor

Rudy Swennen, Department of Earth and Environmental Sciences, K.U. Leuven, Heverlee, Belgium

The book series "Advances in Oil and Gas Exploration & Production" focuses on publishing scientific monographs covering a wide spectrum of topics related to geophysical and geological research within both conventional and unconventional oil and gas systems. These topics are approached from both exploration and production perspectives. The series aims to establish a diverse library of reference works that describe the current state of research on selected themes, such as specific techniques used in the petroleum geoscience industry or regional aspects. Notably, all books in this series are authored and edited by leading experts actively engaged in their respective fields.

***

The "Advances in Oil and Gas Exploration & Production" series encompasses single and multi-authored books, edited volumes, as well as Conference Proceedings. You can obtain a Book Proposal Form from our website or request one directly from the Publishing Editor, Qiao Shu, at qiao.shu@springernature.com.

Shirui Wang • Wenyi Hu • Xuqing Wu •
Jiefu Chen

Deep Learning for Seismic Data Enhancement and Representation

Springer

Shirui Wang
Electrical and Computer Engineering
University of Houston
Houston, TX, USA

Xuqing Wu
Information Science Technology
University of Houston
Houston, TX, USA

Wenyi Hu
SLB
Houston, TX, USA

Jiefu Chen
Electrical and Computer Engineering
University of Houston
Houston, TX, USA

ISSN 2509-372X ISSN 2509-3738 (electronic)
Advances in Oil and Gas Exploration & Production
ISBN 978-3-031-75744-0 ISBN 978-3-031-75745-7 (eBook)
https://doi.org/10.1007/978-3-031-75745-7

This Springer imprint is published by the registered company Springer Nature Switzerland AG
The registered company address is: Gewerbestrasse 11, 6330 Cham, Switzerland

To my beloved family, whose unconditional love and support have been my constant source of strength, guidance, and inspiration.

Preface

In the ever-evolving field of geophysics, the pursuit of more accurate and detailed subsurface imaging has led to the development of increasingly sophisticated methods for seismic data processing and analysis. This work stands as a testament to such advancements, aiming to present an inspiring exploration of novel approaches for seismic data enhancement and representation. At its core, this study is driven by challenges posed by the efficiency-cost trade-off that characterizes modern seismic acquisition strategies.

Central to this work are four pivotal areas: low-frequency seismic data extrapolation, seismic data deblending, seismic data interpolation, and seismic data compression. Each area is a key element in the quest of enhanced seismic data quality and operational efficiency. The advent of simultaneous source acquisition, or blended acquisition, exemplifies the industry's shift toward cost-effective and efficient methodologies. Yet, the complexity of this strategy necessitates sophisticated deblending algorithms to extract clear signals from the interfering noise. The advancement of low-frequency extrapolation for bandwidth extension, the quest for higher data resolution through interpolation, and the imperative for effective data compression in the face of burgeoning data volumes highlight the need for cutting-edge solutions. These solutions navigate the constraints of traditional seismic acquisition techniques, offering a comprehensive approach to overcoming the challenges in seismic data processing.

At the core of this narrative is the transformative potential of deep learning in seismic data processing. Venturing beyond the realms of conventional supervised learning, this book delves into the untapped potential of transfer learning and self-supervised learning schemes, specifically designed to address the prevalent issues of limited labeled training datasets and overfitting. It meticulously tailors novel networks and algorithms to tackle the nuanced challenges of seismic data, not only demonstrating the practicality of deep learning techniques but also integrating the sophisticated physics underlying seismic exploration.

Through rigorous testing against both synthetic and field datasets, the methodologies proposed in this book are validated for their effectiveness and adaptability. This preface invites readers into a narrative that not only chronicles advancements in seismic exploration technologies but also emphasizes the critical role of deep learning in surmounting the field's most pressing challenges.

As you embark on this journey, may the insights and innovations contained within these pages inspire and equip you with the knowledge to navigate the complexities of modern seismic data processing, propelled by the promise of deep learning as catalysts for the future of seismic exploration.

Houston, TX, USA
February 2024

Shirui Wang

Acknowledgments

In crafting this book, I've been fortunate to have the support and encouragement of many remarkable individuals. This journey, which began as an academic endeavor, has evolved into something far greater, thanks in no small part to their contributions. It is with immense gratitude that I acknowledge their roles in this process.

First and foremost, I would like to express my deepest gratitude to my academic advisor Dr. Jiefu Chen, for his invaluable guidance and mentorship throughout my research journey, as well as his support and sponsorship for this book. Dr. Chen not only introduced me to the nuances of this industry but also fostered my academic pursuits and significantly influenced my professional trajectory. He has been a constant source of motivation, helping me to discover my interests and capabilities in this field. I am also grateful for his unwavering belief, encouragement, and patience during every obstacle we encountered. He has provided me with endless support and precious advice in every important moment during these years. His dedication and commitment helped me grow both as a researcher and as an individual.

I am also deeply thankful for my co-advisor Dr. Xuqing Wu, who has always brought his expertise, knowledge, kindness, and patience to our every interaction. His guidance and support have been instrumental in helping me to develop the skills and knowledge necessary for every achievement throughout my academic pursuit. I feel privileged to have had the opportunity to work with him and to learn from his extensive experience in this industry. His hands-on mentorship has been transformative for me, taking me from a new and inexperienced student to a confident researcher. His contribution has been pivotal to my journey.

I would like to acknowledge and extend my sincere appreciation to Dr. Wenyi Hu. We first met in 2018 and then work more closely since 2020 on a project. He has been more than a coworker, but also a mentor and teacher, selflessly and endlessly helping me during and even after the project. His expertise and willingness to share his knowledge have been indispensable in shaping my research and professional development.

Additionally, I acknowledge Dr. Yuchen Jin for his invaluable contribution to this book, enriching it with his insights and expertise.

Above all, I dedicate this book to my parents, Wanfu Wang and Simin Fu. Their unwavering support and belief in the value of education have been the foundation of my pursuits. They fostered an environment of growth, instilling virtues of hard work, perseverance, and determination. It is my earnest hope that this book makes them proud, a reflection of their love, support, and the sacrifices they have made. Their guidance has been my constant light, for which I am forever grateful.

Contents

Acronyms

AE	Arithmetic encoding
AD	Arithmetic decoding
AGC	Automatic gain control
AVO	Amplitude versus offset
BPP	Bits per pixel
BTN	Blind-trace neural network
BTN-SS	Blind-trace neural network with spectrum suppression
BWE	Bandwidth extension
CAE	Convolutional autoencoder
CNN	Convolutional neural network
DC	Decoding
DCT	Discrete cosine transform
DFT	Discrete Fourier transform
DIDC	Dip-informed deformable convolution
DINN	Deep-informed neural network
DNN	Deep neural network
DSP	Deep seismic prior
DWT	Discrete wavelet transform
EC	Encoding
ELM	Extreme learning machine
FFT	Fast Fourier transform
FWI	Full waveform inversion
f-k	Frequency-spectrum
f-x	Frequency-spatial
GPU	Graphics processing unit
HEVC	High efficiency video coding
HF	High frequency
IFFT	Inverse fast Fourier transform
ISTA	Iterative shrinkage threshold algorithm
JPEG	Joint photographic experts group
LF	Low frequency
LPF	Low-pass filter
LSTM	Long short-term memory
MAE	Mean absolute error
MSE	Mean squared error
MSSA	Multichannel singular spectrum analysis
MWNI	Minimum weighted norm interpolation

PEF	Prediction-error filter
POCS	Projection onto convex set
PSNR	Peak signal-to-noise ratio
PTL	Progressive transfer learning
PWD	Plane wave destruction
ReLU	Rectified linear unit
RNN	Recurrent neural network
SNR	Signal-to-noise ratio
SSA	Singular spectra analysis
ST	Seislet transform
SVM	Support vector machine
t-x	Time-spatial
VAE	Variational autoencoder
WGAN	Wasserstein generative adversarial network
WHT	Walsh-Hadamard transform
WT	Wavelet transform
1D	One dimensional
2D	Two dimensional
3D	Three dimensional

Introduction

1

1.1 Balancing Acquisition Efficiency and Data Quality

Seismic technology plays a crucial role in geophysical exploration, serving as a foundation for understanding the Earth's subsurface structures and properties (Yilmaz, 2001; Kearey et al., 2002). The quality of seismic data acquisition systems, combined with effective seismic processing algorithms, is essential to producing accurate and detailed seismic gathers, which are indispensable for subsequent processing, imaging, interpretation, and analysis procedures (Baysal et al., 1983; Scales, 1995; Sirgue and Pratt, 2004; Biondi, 2006; Virieux and Operto, 2009; Liu and Gu, 2012). As depicted in Fig. 1.1, conventional seismic acquisition typically involves firing a single source at a time, with the resulting seismic waves recorded by receivers deployed at the surface. To ensure the cleanness of each shot and to prevent source interference, this process is often repeated with relatively long time intervals between each firing (Berkhout, 2008). However, numerous factors can influence the quality of the recorded data, such as source and receiver density, technical equipment, subsurface complexity, acquisition parameters, environmental conditions, and restrictions, among others. Consequently, seismic data often exhibits density variations across dimensions, with one dimension being dense while another is sparse. As a result, achieving high-quality seismic acquisition typically demands increased manual labor and financial resources, often compromising efficiency. The trade-off between the quality and efficiency of seismic acquisition is always a major concern.

Addressing the challenge of balancing data quality, acquisition efficiency, and cost minimization is crucial for the effectiveness of geophysical exploration. To address the challenge of balancing data quality, acquisition efficiency, and cost minimization, a range of technologies and algorithms have been developed that can be broadly categorized into three groups in this book as illustrated in Fig. 1.2, i.e. (1) Data enhancement algorithms (2) Acquisition optimization techniques, and (3) Data management strategies.

Data enhancement algorithms aim at directly improving the acquired seismic data quality and transforming the raw seismic data into a more interpretable and usable form. These algorithms encompass a range of techniques, including but not limited to denoising, deconvolution, interpolation, and bandwidth extension. A broad definition of seismic denoising includes the removal of various unwanted signals or noises, preservation of essential features and patterns, and improvement of the signal-to-noise ratio (SNR) of the original seismic gathers. Seismic noises can be both incoherent and coherent. Incoherent noises (Canales, 1984) often exhibit no specific patterns or structures and differ from the desired signal in time-spatial (Abma and Claerbout, 1995; Liu

S. Wang et al., *Deep Learning for Seismic Data Enhancement and Representation*, Advances in Oil and Gas Exploration & Production, https://doi.org/10.1007/978-3-031-75745-7_1

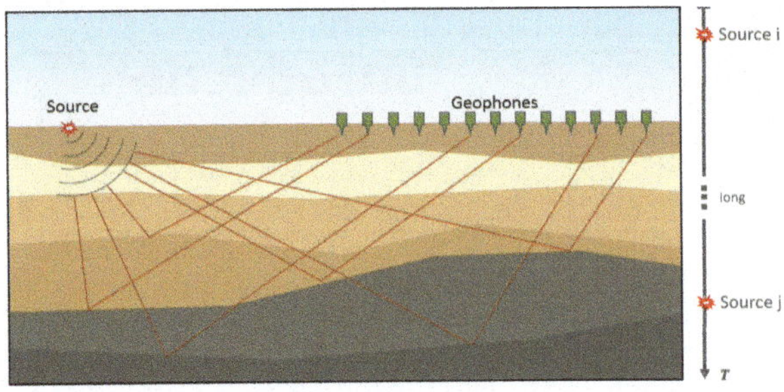

Fig. 1.1 Illustration of land-based seismic acquisition. A source is fired at a location and the seismic signals are recorded by the receivers. The time intervals between sources are relatively long to ensure the cleanness of the seismic record

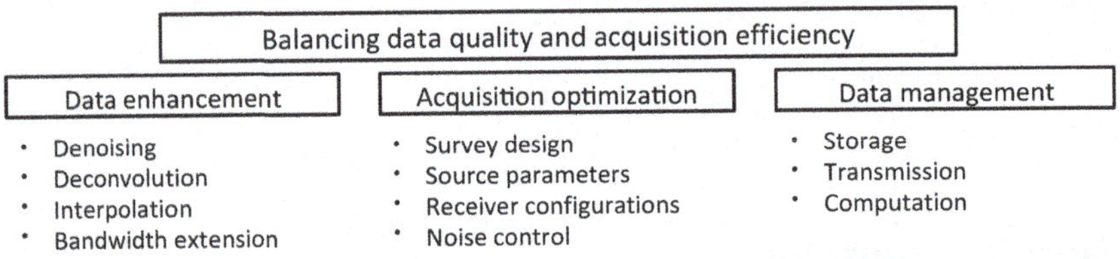

Fig. 1.2 Aspects for balancing the quality-efficiency trade-off

et al., 2009) or certain transformed domains (Hennenfent and Herrmann, 2006; Fomel and Liu, 2010). On the other hand, coherent signals such as ground roll and multiples exhibit distinct spatial or temporal continuity and can be suppressed by filtering or transform-based methods (Deighan and Watts, 1997; Liu, 1999; Weglein, 1999; Weglein et al., 2011). Deconvolution (Sacchi, 1997; Yuan et al., 2017) addresses an inverse problem to resolve overlapping seismic events and remove the effects of the seismic wavelet, thereby enhancing the temporal resolution and producing a clearer image of the subsurface structures. Seismic data interpolation is another vital technique for seismic data enhancement (Spitz, 1991; Liu and Sacchi, 2004; Naghizadeh and Sacchi, 2010). Uniformly sampled seismic data is uncommon in practical seismic acquisition cases. To address defected traces, accurate interpolation algorithms are required to learn the seismic structure from the undecimated record and reconstruct the missing gaps (Abma and Kabir, 2006). Additionally, even if the data is complete and clean, interpolation algorithms are needed when data is aliased due to sparse sampling sensor grids (Gan et al., 2015). Apart from spatial resolution, the frequency spectrum of seismic data can be also enhanced through bandwidth extension, which can improve the stability and robustness of the inversion process (Smith et al., 2008; Xie, 2013).

Acquisition optimization refers to the process of designing and planning seismic data acquisition surveys with the goal of maximizing data quality while minimizing cost and operational complexity. Traditional seismic acquisition emphasizes the cleanness of each shot, necessitating ample time intervals to avoid crosstalk between sources. In recent years, a more productive simultaneous source acquisition strategy has been explored and proven practically feasible. The key concept of simultaneous shooting involves firing multiple source groups synchronously or within shorter intervals, allowing for interfered source responses (Beasley et al., 1998; Bagaini, 2006; Berkhout, 2008). The blended signals then need to be separated as if they were collected individually, using a post-processing algorithm known as seismic deblending.

Data management, although not as extensively discussed in the current literature as data enhancement and acquisition optimization, is increasingly relevant given the growing volume of seismic data. As advanced seismic acquisition techniques and high-resolution sensors generate ever-larger datasets, the need for efficient and effective data management becomes paramount. Seismic data compression plays a critical role in addressing this need by reducing data size without sacrificing essential information (Wood, 1974; Spanias et al., 1991; Helal et al., 2021). Efficient compression techniques enable more effective storage, transfer, and processing of seismic data, ultimately contributing to a streamlined and cost-effective geophysical exploration workflow.

The quality-efficiency trade-off balance in seismic data acquisition and processing has been a topic of discussion for many years. The core motivation behind addressing this challenge is to develop effective and efficient data processing algorithms that can find the optimal representation for seismic data, reconstruct the contaminated signal, and enrich valuable data components.

1.2 Conventional Methods for Seismic Data Processing

Conventional seismic data processing algorithms are designed in various ways, taking into account the nature of the data, task objectives, and specific challenges and requirements involved. While some methods can fit in distinct category, hybrid methods that combine multiple approaches are also prevalent.

Statistical and adaptive filtering represents a major category for seismic data processing. By leveraging the statistical and pattern uniqueness of seismic data, different filters are constructed to suppress noises (Liu et al., 2006, 2009) and enhance data patterns (Bednar, 1983). Predictive filtering methods focus on modeling, predicting, and separating specific components of seismic data based on past values (Canales, 1984; Chen and Ma, 2014). While many approaches process the data in the original time-space domain, others transform the original seismic gathers into different domains, where the data exhibit specific characteristics and are easier to separate from other components. Common transformations include Fourier transform (Xu et al., 2005; Zwartjes and Gisolf, 2006; Gülünay, 2003), Cosine Transform (Lu and Liu, 2007), Curvelet Transform (Starck et al., 2002; Candes et al., 2006; Neelamani et al., 2008; Naghizadeh and Sacchi, 2010), Radon Transform (Trad et al., 2002), Wavelet Transform (Ma et al., 2010; Ouadfeul and Aliouane, 2014; Liu et al., 2015), and Seislet Transform (Fomel and Liu, 2010; Fomel, 2006; Chen and Fomel, 2018). Seismic data is often assumed sparse and resides in a constraint area in the transformed domain. As a result, sparsity regularization is frequently employed in many transform-based methods to address the ill-posedness issue and ensure the reconstruction of seismic data (Fomel, 2007; Gan et al., 2015; Bai and Wu, 2019; Tian et al., 2020). Other approaches leverage the low-rank nature of seismic signals for seismic data representation and reconstruction, such as compressive sensing (Gan et al., 2016) and singular value decomposition (Lu, 2006; Bekara and Van der Baan, 2007; Kreimer and Sacchi, 2012).

Conventional methods have been developed for decades, achieving remarkable success in advancing seismic data processing. While significant results have been attained, the performance of these methods can be hindered by various factors. A critical aspect of filtering-based methods is defining a suitable adaptive filter that can effectively describe the seismic structure and differentiate it from other components. Additionally, many approaches rely on specific regularization and constraints, necessitating meticulous hyperparameter tuning by experienced domain experts. This imposes barriers to automatic and efficient processing workflows.

1.3 Deep Learning for Seismic Data Processing

In recent years, deep learning has seen its fast growth and increased applications for subsurface exploration (Yu and Ma, 2021). The utilization of convolutional neural networks (CNNs) and recurrent neural networks (RNNs) has revolutionized seismic data processing, enabling more accurate imaging and interpretation of subsurface features. These cutting-edge deep learning techniques have been successfully applied to a wide range of interpretation tasks such as event detection (Ross et al., 2018; Mousavi et al., 2019), channel detection (Pham et al., 2019), fault detection (Zhao and Mukhopadhyay, 2018; Xiong et al., 2018), first break picking (Yuan et al., 2019, 2020). Moreover, deep learning has demonstrated remarkable achievements in subsurface imaging tasks such as tomography (Araya-Polo et al., 2018), migration (Zhang et al., 2021) and full waveform inversion (FWI) (Jin et al., 2018; Hu et al., 2021).

Deep learning has also shown its power for seismic data enhancement and representation. For example, deep convolutional neural networks have been constructed for various noise attenuation tasks, including random noise (Jin et al., 2018; Zhu et al., 2019; Yu et al., 2019), coherent noise (Dong et al., 2020), multiple (Wang et al., 2022), ground roll (Kaur et al., 2020; Pham and Li, 2022), and blending noise (Richardson and Feller, 1907). Seismic data quality is also improved through other approaches such as seismic data interpolation (Oliveira et al., 2018), bandwidth extension (Sun and Demanet, 2015), and deconvolution (Gao et al., 2021). The end-to-end deep learning scheme and data-driven system allow the network to learn the hidden information in seismic data, and enable an automatic seismic data processing workflow, in which crucial filter design and hyperparameter tuning problems can be alleviated or avoided.

Despite the remarkable achievements of deep learning in seismic data processing, challenges still exist. The field-specific nature of seismic data acquisition poses a significant challenge, as there is always a large distributional difference between different field datasets due to subsurface complexity, acquisition configuration, source parameters, and environmental noises. Therefore, it is hard to find an optimal training dataset that can guarantee the generalization capability of the trained network. Although data augmentation can mitigate this problem to some extent and methods have been proposed to generate synthetic data which is enough "real" (Wu et al., 2019), it is still not realistic for many tasks to have a synthetic dataset that satisfies the diversity and realism for bridging the distributional gap between training and target datasets.

To ensure the stability of deep learning approaches and reduce their reliance on prepared "input and truth" training pairs, self-supervised learning schemes have emerged as a promising solution that is preferred in many cases. Self-supervised learning for seismic data processing assumes that only the target observation, such as contaminated data for denoising, band-limited data for bandwidth extension, or decimated or under-sampled components for interpolation, is accessible. In this approach, the deep neural network is trained directly using the target dataset without relying on other pre-labeled synthetic or field datasets. Although self-supervised learning has gained more attention in recent years, there is still limited literature exploring its application in seismic data processing.

1.4 Content and Layout of the Book

This book focuses on four major topics related to seismic data enhancement and representation, aimed at addressing the quality-efficiency trade-off in seismic data acquisition and processing. First, we provide an approach for reconstructing the full-bandwidth seismic data from band-limited records. Second, from the perspective of acquisition optimization, we address the significant challenge of deblending in simultaneous source acquisition. Third, we explore self-supervised approaches for seismic data interpolation, which is essential for data enhancement. We categorize the interpolation task into irregular and regular scenarios and address the unique challenges accordingly. Finally, we discuss the deep learning-based seismic data compression approach to address the cost issue associated with seismic data storage, transmission, and computation. The book is divided into individual chapters, each addressing a different task or challenge. Layout of the book is as follows.

Chapter 2 of this book concentrates on extrapolating vital low-frequency information from high-frequency data with the aid of deep learning. We propose a progressive transfer learning approach to overcome the difficulty of acquiring training data for field cases, thus facilitating the low-frequency data extrapolation with limited and non-representative training data. The proposed workflow jointly solves the physics-driven FWI and data-driven deep neural network training and inferencing, thus progressively correcting the bias of the neural network prediction caused by the non-representative nature of the initial training data by implicitly incorporating the physics constraints into the network training. Our method could be flexibly combined with various deep-learning approaches and has been validated by different experimental cases.

Chapter 3 of this book focuses on the seismic deblending problem. It begins by introducing the definition and problem formulation of simultaneous source acquisition, which leads to the need for deblending. The progressive transfer learning method is deployed to address the generalization issue of supervised learning. We then discuss the limitations of supervised learning schemes for seismic data processing. Then, a self-supervised learning approach is proposed to solve the deblending problem. Three different blending schemes are explored in this book, each of which introduces varying noise levels and complexities. The proposed method is evaluated on different datasets and compared with the current literature to demonstrate its effectiveness and superiority.

In Chap. 4, a self-supervised seismic data interpolation method is introduced. We explore and describe the obstacle of self-supervised learning for seismic data interpolation in the current literature, and propose a novel and effective solution to it. The seismic data interpolation task is categorized into regular and irregular scenarios, and we address the challenges in each case with tailored solutions. The chapter presents detailed experiments designed to evaluate the proposed method's effectiveness.

Chapter 5 continues the discussion of the seismic data interpolation problem, focusing specifically on the regular decimation case. We address the challenge of regular interpolation discussed in the previous chapter and propose a solution. An anti-aliasing self-supervised seismic interpolation algorithm is constructed with a slope-informed deep neural network. We show that the network is tailored for seismic data processing and effectively leverages the physics information of seismic gathers. Both synthetic and field datasets are utilized to evaluate the proposed method.

Then, Chap. 6 presents a deep learning application for the seismic data compression. We show that both pre-migration and post-migration seismic data can be effectively compressed with a high compression ratio and subsequently decompressed with high reconstruction fidelity.

Finally, This book is concluded in Chap. 7.

References

Abma, R., Claerbout, J.: Lateral prediction for noise attenuation by t-x and f-x techniques. Geophysics **60**(6), 1887–1896 (1995)

Abma, R., Kabir, N.: 3D interpolation of irregular data with a POCS algorithm. Geophysics **71**(6), E91–E97 (2006)

Araya-Polo, M., Jennings, J., Adler, A., Dahlke, T.: Deep-learning tomography. Lead. Edge **37**(1), 58–66 (2018)

Bagaini, C.: Overview of simultaneous vibroseis acquisition methods. In: 2006 SEG Annual Meeting. OnePetro, Richardson (2006)

Bai, M., Wu, J.: Iterative sparse deconvolution using seislet-domain constraint. J. Seismic Explor. **28**(1), 73–88 (2019)

Baysal, E., Kosloff, D.D., Sherwood, J.W.: Reverse time migration. Geophysics **48**(11), 1514–1524 (1983)

Beasley, C.J., Chambers, R.E., Jiang, Z.: A new look at simultaneous sources. In: SEG Technical Program Expanded Abstracts 1998, pp. 133–135. Society of Exploration Geophysicists, Houston (1998)

Bednar, J.B.: Applications of median filtering to deconvolution, pulse estimation, and statistical editing of seismic data. Geophysics **48**(12), 1598–1610 (1983)

Bekara, M., Van der Baan, M.: Local singular value decomposition for signal enhancement of seismic data. Geophysics **72**(2), V59–V65 (2007)

Berkhout, A.E.: Changing the mindset in seismic data acquisition. Lead. Edge **27**(7), 924–938 (2008)

Biondi, B.L.: 3D seismic imaging. Society of Exploration Geophysicists, Houston (2006)

Canales, L.L.: Random noise reduction. In: SEG Technical Program Expanded Abstracts 1984, pp. 525–527. Society of Exploration Geophysicists, Houston (1984)

Candes, E., Demanet, L., Donoho, D., Ying, L.: Fast discrete curvelet transforms. Multiscale Model. Simul. **5**(3), 861–899 (2006)

Chen, Y., Fomel, S.: EMD-seislet transform. Geophysics **83**(1), A27–A32 (2018)

Chen, Y., Ma, J.: Random noise attenuation by fx empirical-mode decomposition predictive filtering. Geophysics **79**(3), V81–V91 (2014)

Deighan, A.J., Watts, D.R.: Ground-roll suppression using the wavelet transform. Geophysics **62**(6), 1896–1903 (1997)

Dong, X., Li, Y., Zhong, T., Wu, N., Wang, H.: Random and coherent noise suppression in DAS-VSP data by using a supervised deep learning method. IEEE Geosci. Remote Sens. Lett. **19**, 1–5 (2020)

Fomel, S.: Towards the seislet transform. In: 2006 SEG Annual Meeting. OnePetro, Richardson (2006)

Fomel, S.: Shaping regularization in geophysical-estimation problems. Geophysics **72**(2), R29–R36 (2007)

Fomel, S., Liu, Y.: Seislet transform and seislet frame. Geophysics **75**(3), V25–V38 (2010)

Gan, S., Wang, S., Chen, Y., Zhang, Y., Jin, Z.: Dealiased seismic data interpolation using seislet transform with low-frequency constraint. IEEE Geosci. Remote Sens. Lett. **12**(10), 2150–2154 (2015)

Gan, S., Wang, S., Chen, Y., Chen, X., Huang, W., Chen, H.: Compressive sensing for seismic data reconstruction via fast projection onto convex sets based on seislet transform. J. Appl. Geophys. **130**, 194–208 (2016)

Gao, Z., Hu, S., Li, C., Chen, H., Jiang, X., Pan, Z., Gao, J., Xu, Z.: A deep-learning-based generalized convolutional model for seismic data and its application in seismic deconvolution. IEEE Trans. Geosci. Remote Sens. **60**, 1–17 (2021)

Gülünay, N.: Seismic trace interpolation in the fourier transform domain. Geophysics **68**(1), 355–369 (2003)

Helal, E.B., Saad, O.M., Hafez, A.G., Chen, Y., Dousoky, G.M.: Seismic data compression using deep learning. IEEE Access **9**, 58161–58169 (2021)

Hennenfent, G., Herrmann, F.J.: Seismic denoising with nonuniformly sampled curvelets. Comput. Sci. Eng. **8**(3), 16–25 (2006)

Hu, W., Jin, Y., Wu, X., Chen, J.: Progressive transfer learning for low-frequency data prediction in full-waveform inversion. Geophysics **86**(4), R369–R382 (2021)

Jin, Y., Hu, W., Wu, X., Chen, J.: Learn low wavenumber information in FWI via deep inception based convolutional networks. In: 2018 SEG International Exposition and Annual Meeting. OnePetro, Richardson (2018)

Jin, Y., Wu, X., Chen, J., Han, Z., Hu, W.: Seismic data denoising by deep-residual networks. In: SEG Technical Program Expanded Abstracts 2018, pp. 4593–4597. Society of Exploration Geophysicists, Houston (2018)

Kaur, H., Fomel, S., Pham, N.: Seismic ground-roll noise attenuation using deep learning. Geophys. Prospect. **68**(7), 2064–2077 (2020)

Kearey, P., Brooks, M., Hill, I.: An Ntroduction to Geophysical Exploration, vol. 4. Wiley, Hoboken (2002)

Kreimer, N., Sacchi, M.D.: A tensor higher-order singular value decomposition for prestack seismic data noise reduction and interpolation. Geophysics **77**(3), V113–V122 (2012)

Liu, X.: Ground roll suppression using the Karhunen-Loeve transform. Geophysics **64**(2), 564–566 (1999)

Liu, Q., Gu, Y.: Seismic imaging: from classical to adjoint tomography. Tectonophysics **566**, 31–66 (2012)

Liu, B., Sacchi, M.D.: Minimum weighted norm interpolation of seismic records. Geophysics **69**(6), 1560–1568 (2004)

Liu, C., Liu, Y., Yang, B., Wang, D., Sun, J.: A 2D multistage median filter to reduce random seismic noise. Geophysics **71**(5), V105–V110 (2006)

Liu, Y., Liu, C., Wang, D.: A 1D time-varying median filter for seismic random, spike-like noise elimination. Geophysics **74**(1), V17–V24 (2009)

Liu, W., Cao, S., Chen, Y.: Seismic time-frequency analysis via empirical wavelet transform. IEEE Geosci. Remote Sens. Lett. **13**(1), 28–32 (2015)

Lu, W.: Adaptive noise attenuation of seismic images based on singular value decomposition and texture direction detection. J. Geophys. Eng. **3**(1), 28–34 (2006)

Lu, W.k., Liu, J.: Random noise suppression based on discrete cosine transform. In: 2007 SEG Annual Meeting. OnePetro, Richardson (2007)

Ma, J., Plonka, G., Chauris, H.: A new sparse representation of seismic data using adaptive easy-path wavelet transform. IEEE Geosci. Remote Sens. Lett. **7**(3), 540–544 (2010)

Mousavi, S.M., Zhu, W., Sheng, Y., Beroza, G.C.: CRED: a deep residual network of convolutional and recurrent units for earthquake signal detection. Sci. Rep. **9**(1), 1–14 (2019)

Naghizadeh, M., Sacchi, M.D.: Beyond alias hierarchical scale curvelet interpolation of regularly and irregularly sampled seismic data. Geophysics **75**(6), WB189–WB202 (2010)

Neelamani, R., Baumstein, A.I., Gillard, D.G., Hadidi, M.T., Soroka, W.L.: Coherent and random noise attenuation using the curvelet transform. Lead. Edge **27**(2), 240–248 (2008)

Oliveira, D.A., Ferreira, R.S., Silva, R., Brazil, E.V.: Interpolating seismic data with conditional generative adversarial networks. IEEE Geosci. Remote Sens. Lett. **15**(12), 1952–1956 (2018)

Ouadfeul, S.A., Aliouane, L.: Random seismic noise attenuation data using the discrete and the continuous wavelet transforms. Arabian J. Geosci. **7**, 2531–2537 (2014)

Pham, N., Li, W.: Physics-constrained deep learning for ground roll attenuation. Geophysics **87**(1), V15–V27 (2022)

Pham, N., Fomel, S., Dunlap, D.: Automatic channel detection using deep learning. Interpretation **7**(3), SE43–SE50 (2019)

Richardson, A., Feller, C.: Seismic data denoising and deblending using deep learning. arXiv:1907.01497 (2019)

Ross, Z.E., Meier, M.A., Hauksson, E., Heaton, T.H.: Generalized seismic phase detection with deep learning. Bullet. Seismol. Soc. Am. **108**(5A), 2894–2901 (2018)

Sacchi, M.D.: Reweighting strategies in seismic deconvolution. Geophys. J. Int. **129**(3), 651–656 (1997)

Scales, J.A.: Theory of Seismic Imaging, vol. 2. Springer, Berlin (1995)

Sirgue, L., Pratt, R.G.: Efficient waveform inversion and imaging: a strategy for selecting temporal frequencies. Geophysics **69**(1), 231–248 (2004)

Smith, M., Perry, G., Stein, J., Bertrand, A., Yu, G.: Extending seismic bandwidth using the continuous wavelet transform. First Break **26**(6) (2008)

Spanias, A.S., Jonsson, S.B., Stearns, S.D.: Transform methods for seismic data compression. IEEE Trans. Geosci. Remote Sens. **29**(3), 407–416 (1991)

Spitz, S.: Seismic trace interpolation in the FX domain. Geophysics **56**(6), 785–794 (1991)

Starck, J.L., Candès, E.J., Donoho, D.L.: The curvelet transform for image denoising. IEEE Trans. Image Process. **11**(6), 670–684 (2002)

Sun, H., Demanet, L.: Low frequency extrapolation with deep learning. In: SEG Technical Program Expanded Abstracts 2018, pp. 2011–2015. Society of Exploration Geophysicists, Houston (2018)

Tian, K., Liu, L., Wang, C., Liang, H., Li, G., Zhu, L.: Sparse constrained multi-source least-squares reverse-time migration using seislet transform. In: SEG Technical Program Expanded Abstracts 2020, pp. 3078–3082. Society of Exploration Geophysicists, Houston (2020)

Trad, D.O., Ulrych, T.J., Sacchi, M.D.: Accurate interpolation with high-resolution time-variant radon transforms. Geophysics **67**(2), 644–656 (2002)

Virieux, J., Operto, S.: An overview of full-waveform inversion in exploration geophysics. Geophysics **74**(6), WCC1–WCC26 (2009)

Wang, K., Hu, T., Wang, S., Wei, J.: Seismic multiple suppression based on a deep neural network method for marine data. Geophysics **87**(4), V341–V365 (2022)

Weglein, A.B.: Multiple attenuation: an overview of recent advances and the road ahead (1999). Lead. Edge **18**(1), 40–44 (1999)

Weglein, A.B., Hsu, S.Y., Terenghi, P., Li, X., Stolt, R.H.: Multiple attenuation: recent advances and the road ahead (2011). Lead. Edge **30**(8), 864–875 (2011)

Wood, L.C.: Seismic data compression methods. Geophysics **39**(4), 499–525 (1974)

Wu, X., Liang, L., Shi, Y., Fomel, S.: FaultSeg3D: using synthetic data sets to train an end-to-end convolutional neural network for 3d seismic fault segmentation. Geophysics **84**(3), IM35–IM45 (2019)

Xie, X.B.: Recover certain low-frequency information for full waveform inversion. In: 2013 SEG Annual Meeting. OnePetro, Utrecht (2013)

Xiong, W., Ji, X., Ma, Y., Wang, Y., AlBinHassan, N.M., Ali, M.N., Luo, Y.: Seismic fault detection with convolutional neural network. Geophysics **83**(5), O97–O103 (2018)

Xu, S., Zhang, Y., Pham, D., Lambaré, G.: Antileakage fourier transform for seismic data regularization. Geophysics **70**(4), V87–V95 (2005)

Yilmaz, Ö.: Seismic data analysis: processing, inversion, and interpretation of seismic data. Society of Exploration Geophysicists, Houston (2001)

Yu, S., Ma, J.: Deep learning for geophysics: current and future trends. Rev. Geophys. **59**(3), e2021RG000742 (2021)

Yu, S., Ma, J., Wang, W.: Deep learning for denoising. Geophysics **84**(6), V333–V350 (2019)

Yuan, S., Wang, S., Ma, M., Ji, Y., Deng, L.: Sparse Bayesian learning-based time-variant deconvolution. IEEE Trans. Geosci. Remote Sens. **55**(11), 6182–6194 (2017)

Yuan, P., Hu, W., Wu, X., Chen, J., Van Nguyen, H.: First arrival picking using U-net with lovasz loss and nearest point picking method. In: SEG International Exposition and Annual Meeting. OnePetro, Richardson (2019)

Yuan, P., Wang, S., Hu, W., Wu, X., Chen, J., Van Nguyen, H.: A robust first-arrival picking workflow using convolutional and recurrent neural networks. Geophysics **85**(5), U109–U119 (2020)

Zhang, W., Gao, J., Jiang, X., Sun, W.: Consistent least-squares reverse time migration using convolutional neural networks. IEEE Trans. Geosci. Remote Sens. **60**, 1–18 (2021)

Zhao, T., Mukhopadhyay, P.: A fault detection workflow using deep learning and image processing. In: 2018 SEG International Exposition and Annual Meeting. OnePetro, Richardson (2018)

Zhu, W., Mousavi, S.M., Beroza, G.C.: Seismic signal denoising and decomposition using deep neural networks. IEEE Trans. Geosci. Remote Sens. **57**(11), 9476–9488 (2019)

Zwartjes, P., Gisolf, A.: Fourier reconstruction of marine-streamer data in four spatial coordinates. Geophysics **71**(6), V171–V186 (2006)

2 Full Waveform Inversion with Low-Frequency Extrapolation

2.1 Introduction to Full Waveform Inversion

Full waveform inversion (FWI) is an advanced seismic processing technology for reconstructing high-resolution, subsurface geophysical models utilizing entire waveform information, which differentiates it from conventional velocity tomography that uses kinematic information only for the purpose of velocity model building. The mathematical foundation that FWI is based on can be formulated as a data-fitting algorithm between the forward synthesis and observations (Virieux and Operto, 2009; Virieux et al., 2017). The forward problem in FWI is essentially based on the wave equation. We refer the readers to the reference (Virieux et al., 2017) for further information on this forward modeling problem. The unknown velocity model is denoted by $\mathbf{v}$. $\mathbf{d}^{(syn)}$ and $\mathbf{d}^{(obs)}$ represent the synthesized data and the observed data, respectively. The general form of the cost function used in FWI can be formulated as

$$C(\mathbf{v}) = \sum_{f=1}^{N_f} \sum_{s=1}^{N_s} \sum_{r=1}^{N_r} \left\| \mathbf{d}_{r,s,f}^{(syn)}(\mathbf{v}) - \mathbf{d}_{r,s,f}^{(obs)} \right\|_2^2, \tag{2.1}$$

where N_r, N_s, N_f represent the number of receivers, sources, and frequency components, respectively. Depending on what specific FWI algorithm is used, the form of the observations $\mathbf{d}$ may vary. Conventionally, FWI is typically performed in either the time-domain (Ha et al., 2015; Gholami et al., 2022), or the frequency domain (Shin and Ho Cha, 2009; Kim et al., 2013). In field data applications, the amplitude information is often unreliable for inversion due to an unknown density model, inaccurate attenuation property, acoustic approximation, low signal-to-noise ratio, and other factors that cannot be taken into account in the FWI forward modeling engine. Some researchers show that the phase-only FWI (Bednar et al., 2007; Maharramov et al., 2017; Sheng et al., 2020) improves the robustness and the performance in those scenarios. We refer the reader to Hu et al. (2018) for further references, where the authors summarize and compare different FWI methods, including the phase-only FWI.

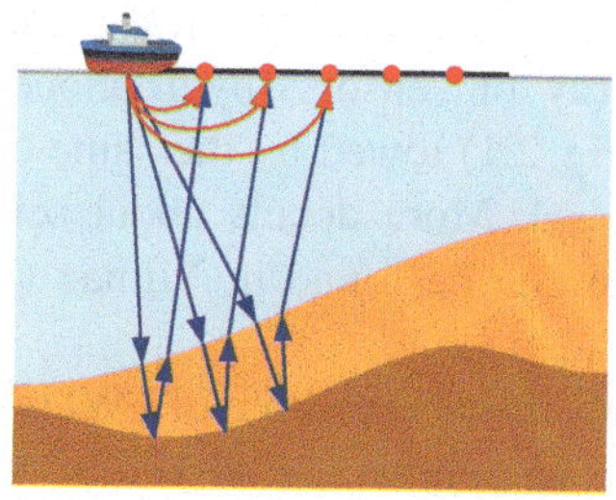

Fig. 2.1 An example of the data acquisition for FWI. The red lines and blue lines represent transmission waves and reflection waves respectively

In Fig. 2.1, we show an example of the marine seismic data acquisition survey. The sources, sometimes also called "shots" that are config-

S. Wang et al., *Deep Learning for Seismic Data Enhancement and Representation*, Advances in Oil and Gas Exploration & Production, https://doi.org/10.1007/978-3-031-75745-7_2

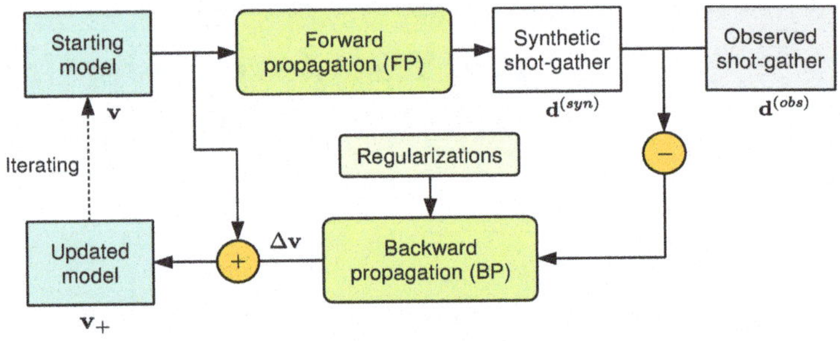

Fig. 2.2 The diagram of the data-fitting algorithm used in FWI

ured by a source wavelet, fire a series of seismic waves as the acquisition survey vessel travels. The time series of the reflected and refracted seismic waves captured by receivers are called "seismic traces". The seismic traces are recorded by the array of sensors (hydrophone streamer cable in Fig. 2.1) towed by the same acquisition survey vessel. More details about seismic data acquisition can be found in Yilmaz (2001) and Dondurur (2018).

The data-fitting algorithm of FWI is illustrated in Fig. 2.2. In general, the algorithm starts from an initial velocity model. For each loop, (1) we use the forward wave propagation equations to synthesize shot gathers from the current velocity model; (2) then we calculate the data residual between the synthetic data and the observations; (3) the data residual, along with the regularization terms, is back-propagated to the subsurface and cross-correlated to the forward wavefield to form the gradient with respect to the velocity model (Asnaashari et al., 2012; Xue et al., 2017; Qiu et al., 2016); (4) the obtained gradient is used for calculating the update of the velocity model; (5) after that, the algorithm returns to (1) and starts a new iteration until the model converges. Although the forward propagation of FWI is generally highly efficient, the solution space of 2D or 3D FWI is usually very large, and thus, the inversion process in an FWI workflow is memory and time-consuming. Most FWI algorithms are Hessian-free methods to avoid calculating matrix inverse and limit the memory use or the computational cost (Pan et al., 2015, 2017). Instead, the velocity model updates are usually estimated by the steepest descent method (Yan, 2017; Ren, 2022), the conjugate gradient method (Pan et al., 2017; He and Han, 2018) or other proximal algorithms (Aghamiry et al., 2021; Fu et al., 2020).

2.1.1 Challenges of Capturing the Low-Frequency and Low-Wavenumber Information in FWI

Mathematically, FWI is an underdetermined inverse problem. Without a good initial velocity model in a kinematic sense and sufficiently low-frequency components to launch an FWI process, the strong nonlinearity nature of this inverse problem may lead the gradient-based search to be easily trapped in local minima, a well-known problem called cycle skipping (Virieux et al., 2017; Hu et al., 2018). When strong reflectors like salt domes and carbonates exist in the geological environment, FWI will become more challenging. Since only a small portion of low-frequency seismic energy is able to penetrate those strong reflectors, it is difficult to reconstruct the sub-salt structures with band-limited data. In Wang et al. (2019), the authors summarize the state-of-art progress of FWI for salt reconstruction and discuss the importance of different frequency components in FWI. In Fig. 2.3, we show an example of FWI inverting multiple frequency components. Starting from an initial model and the low-frequency data, FWI provides a low-resolution model containing the correct low-wavenumber information. This low-resolution velocity model will be subsequently used as the initial model in the next round of FWI, where the higher-frequency data is inverted to resolve the higher-wavenumber components in the subsurface velocity model. The inversion process continues in an interactive manner until the highest-frequency component is inverted. As a comparison, in another scenario where the low-frequency information is lacking, the high-frequency only FWI shown

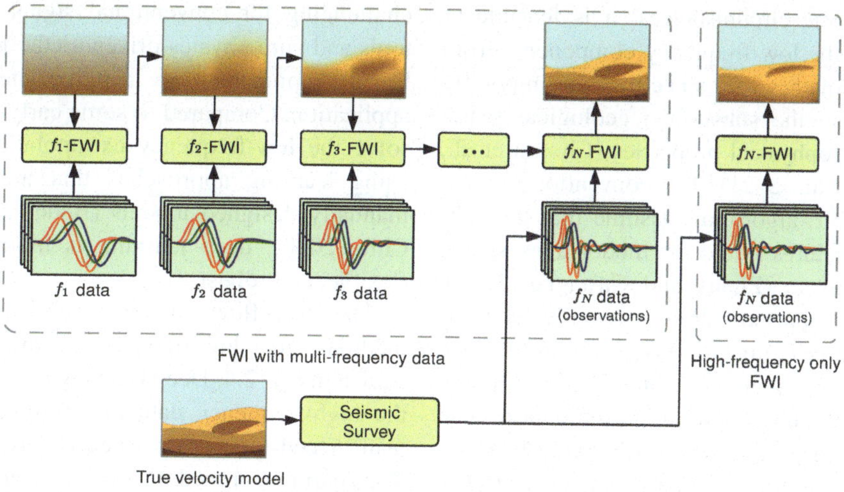

Fig. 2.3 Comparison between multi-frequency FWI and high-frequency only FWI

in Fig. 2.3 often fails to converge to the correct solution. According to Wang et al. (2019), low-frequency solutions contain large-scale structural information of the velocity model and provide a good initial condition for the convergences of the subsequent higher-frequency FWIs. In comparison, high-frequency data carries high-wavenumber information of the subsurface geophysical models and plays an important role in resolving the fine structural features in the subsurface.

Due to the limitation of the seismic acquisition device, the recorded seismic data during seismic acquisition surveys are band-limited. In most seismic datasets, the low-frequency components are noisy and unreliable. The absence of low-frequency components causes FWI to fail to converge to the global minimum. To tackle this problem, previously, researchers proposed various methods to synthesize the low-frequency information or retrieve the low-wavenumber information from the band-limited observations. A typical industrial practice is to develop low-frequency sources. Such expensive sources can provide low-frequency data directly (Brenders et al., 2022). A more economical approach is to solve the low-wavenumber model by algorithms. For example, travel time tomography only focuses on low-wavenumber features and low-resolution model (Alkhalifah and Choi, 2014). In Irabor and Warner (2016), the authors propose reflection FWI, which combines migration and tomography. This method avoids the high-frequency FWI and provides the low-resolution velocity model directly. Another idea is to extract the low-frequency information from the observation by the atom decomposition (Bénar et al., 2007). It is also possible to synthesize "fake" low-frequency data. For instance, the envelop-based inversion (Chi et al., 2014; Liu et al., 2018) uses the envelope of observations to replace the low-frequency data during the FWI. The envelope can also be used to construct adjoint sources containing low-frequency information (Oh and Alkhalifah, 2018). A more direct approach is to extract low-frequency data from high-frequency data explicitly. A popular low-frequency extrapolation method is the sparsity-promoted BWE (Liang et al., 2017; Zhang et al., 2017; Chiu et al., 2018; Zhang and Castagna, 2011; Chopra et al., 2006). There are some other effective low-frequency extrapolation approaches like Fourier-filter-based BWE (Xie, 2013), continuous wavelet transform (Smith et al., 2008), and beat tone extraction (Hu, 2014, 2015). These conventional approaches have achieved limited success, especially when the geological structure could be modeled as a stack of sparse layers (Chopra et al., 2006).

Under some circumstances, it is feasible to reconstruct the low-frequency components from the high-frequency data since they are implicitly connected by the subsurface geological structures and geophysical properties (Chopra et al., 2006; Hu et al., 2021). The conventional trace-by-trace BWE algorithms assume that the subsurface structures could be modeled as sparse reflection layers (Zhang and Castagna, 2011), a reasonable approximation, especially for post-stack subsurface images. When the subsurface features become more geologically complex, and the object of study is pre-migration data instead of seismic images, supervised learning is a better choice. The training dataset of the learning-based BWE can be constructed by simultaneously performing the high-frequency and low-frequency simulation. Supervised learning can extract the complicated and hidden relationship between high-frequency and low-frequency data. In comparison, the sparsity assumption might not be justified when the subsurface structures are complex. The lateral continuity also needs to be taken into account. Therefore, the deep learning-based low-frequency extrapolation methods can serve as better alternatives to conventional BWE algorithms.

2.2 Supervised-Learning-Based Method for Low-Frequency Extrapolation

Retrieving the missing low-frequency information from band-limited data is a challenging nonlinear inverse problem that is solvable only under certain conditions. In Hu et al. (2021), the authors discuss the feasibility of the low-frequency extrapolation in seismic exploration applications with and without the subsurface structural sparsity assumption. In both scenarios, the low-frequency components are extracted from the low-wavenumber structural information implicitly contained in the band-limited data. Because rigorously establishing the implicit relationship among the low-frequency components, the low wavenumber information, and the recorded band-limited seismic data is extremely challenging for conventional signal processing tools and pure physics-driven methods, learning-based approaches are appropriate for this application. Compared to some early attempts to solve the low-frequency extrapolation problem using learning approaches that are based on manually designed models (Bader et al., 2019; Chiu, 2020), deep learning is more powerful, flexible, and robust.

The workflow of the deep-learning-based methods for low-frequency extrapolation is shown in Fig. 2.4. Deep learning approaches map the high-frequency data to low-frequency data using a feed-forward deep neural network (DNN). The learning procedure of such general DNNs is purely data-driven and does not necessarily rely on some specific constraints like sparsity-based regularization because these constraints are believed to be automatically incorporated into the network parameters through the training process as long as the training is unbiased. Benefiting from the data-driven architecture and GPU acceleration, after the training, the DNN could predict the low-frequency data rapidly and accurately (Ovcharenko et al., 2018; Liu et al., 2021). If the training datasets are sufficiently large and diverse with high quality and minimized skew, the DNN could work well with various kinds of complicated testing cases (Ovcharenko et al., 2021b,a; Sun and Demanet, 2019). By tweaking the training dataset properly, the prior information can be fed into the DNN training, thus improving the effectiveness of DNNs in practical cases (Ovcharenko et al., 2019a).

In previous works, the design of deep learning architecture has been deeply explored for the frequency extrapolation of both acoustic data (Ovcharenko et al., 2019a; Sun and Demanet, 2020b; Hu et al., 2021) and elastic data (Sun and Demanet, 2020a, 2021; Ovcharenko et al., 2020). The data arrangement strategies in these works can be classified into three types: trace-by-trace, shot-by-shot, and patch-by-patch. The trace-by-trace arrangement is flexible for data with amplitudes that vary from trace to trace (Sun and Demanet, 2020b,a, 2021; Ovcharenko et al., 2021a; Wang et al., 2020). The shot-by-shot arrangement

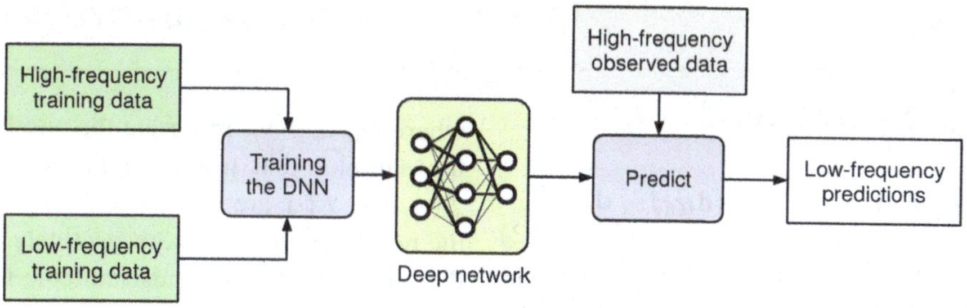

Fig. 2.4 The general workflow of low-frequency extrapolation based on supervised learning

takes more information into consideration as additional constraints during the multi-offset processing (Ovcharenko et al., 2018, 2019b,a, 2021a; Fabien-Ouellet, 2020). The patch-by-patch arrangement is more suitable for large-scale data processing, because it also takes into account the multi-offset information without dramatically increasing the memory usage (Fang et al., 2020; Ovcharenko et al., 2020; Hu et al., 2021, 2020). Researchers have investigated different kinds of DNNs, including the artificial neural network (Ovcharenko et al., 2018), the convolutional neural network (CNN) (Sun and Demanet, 2020b,a, 2021; Ovcharenko et al., 2019a; Fang et al., 2020), InceptionNet (Hu et al., 2021, 2020), U-Net (Ovcharenko et al., 2020), MobileNet (Ovcharenko et al., 2019b), and the recursive convolutional neural network (Fabien-Ouellet, 2020). In Sun and Demanet (2020b), the robustness and the generalization ability of the DNN are studied. In Ovcharenko et al. (2020), the authors compare the performance with and without data pre-processing.

The generalization ability is an important issue for learning-based methods. Since the low-frequency data is absent in recorded field data, the available training datasets of the supervised learning methods are typically synthetic data. The inevitable gap between the synthetic training dataset and the field testing data largely affects the performance of these supervised learning methods. When a training dataset is similar to the testing dataset in terms of multiple seismic attributes, data survey geometry, noise level, noise properties, a geological and geophysical environment where the data are acquired, and many other factors, the supervised learning performance is expected to be improved. A straightforward method for generating a large amount of training data is to extract sub-regions from existing velocity models (Sun and Demanet, 2020b, 2021). Another efficient approach is to generate random velocity models with style transfer DNN (Kazei et al., 2020; Sun et al., 2021). In Ovcharenko et al. (2019b), Hu et al. (2021), Hu et al. (2020), and Jin et al. (2022a), the transfer learning approaches are adopted to improve the similarity between the training data and testing data. In Hu et al. (2020), Wang et al. (2020), and Jin et al. (2022a), the authors propose self-supervised learning methods to achieve good supervised learning performance with a small training dataset.

2.2.1 Training and Testing

Most supervised learning approaches are the variants of using true low-frequency data to train the DNN module. Since the true low-frequency data is unavailable in field surveys, such labeling data needs to be synthesized by the forward simulation. The DNN learns a mapping from the high-frequency data to the low-frequency data. Suppose we have already run the forward simulation on the current velocity model. The ith high-frequency and low-frequency training data provided by the simulation are denoted by $\mathbf{d}_{HF}^{(i)}$, and $\mathbf{d}_{LF}^{(i)}$, respectively. $\mathcal{D}_{\Theta}(\cdot)$ represents the DNN with parameters Θ. The DNN training could be formulated as

$$\arg\min_{\Theta} \frac{1}{N_{tr}} \sum_{i=1}^{N_{tr}} \mathcal{L}\left(\mathbf{d}_{HF}^{(i)}, \mathbf{d}_{LF}^{(i)}\right),$$
$$\mathcal{L}\left(\mathbf{d}_{HF}, \mathbf{d}_{LF}\right) = \|\mathcal{D}_{\Theta}\left(\mathbf{d}_{HF}\right) - \mathbf{d}_{LF}\|_2^2. \tag{2.2}$$

where N_{tr} represent the number of training samples. $\mathcal{L}(\cdot)$ is the mean squared error (MSE) between the DNN predictions and low-frequency labels. The objective function (2.2) is suitable for different types of data acquisitions. For example, when the training data is gathered trace-by-trace, N_{tr} represents the trace numbers. If the data is collected shot-by-shot, N_{tr} represents the shot numbers. Equation (2.2) is also suitable for representing both the frequency-domain and time-domain predictions. In other words, the traces like $\mathbf{d}_{HF}^{(i)}$ can be either a phase component, frequency component or a time series trace.

After training the DNN, we acquire the prediction by feeding the observed band-limited data $\mathbf{d}^{(i)}$ to the DNN directly. The low-frequency data is formulated as

$$\hat{\mathbf{d}}_{\text{LF}} = \Psi\left(\mathcal{D}_{\Theta}\left(\mathbf{d}\right)\right). \tag{2.3}$$

where the function $\Psi(\cdot)$ is a low-pass filter.

2.2.2 Regularization for Training the Time-Domain DNN

When the DNN is designed for accepting and predicting data in the time domain, even if the labeling data is limited by a low-pass filter, the DNN prediction is not granted as low-frequency-only. In other words, the predicted data may contain high-frequency components, which is called "high-frequency leakage." To mitigate this issue, the objective function in (2.2) can be modified as

$$\arg\min_{\Theta} \frac{1}{N_{tr}} \sum_{i=1}^{N_{tr}} \mathcal{L}\left(\mathbf{d}_{HF}^{(i)}, \mathbf{d}_{LF}^{(i)}\right)$$
$$+ \frac{\lambda/2}{N_{tr}} \sum_{i=1}^{N_{tr}} \mathcal{R}\left(\mathbf{d}_{HF}^{(i)}\right) + \frac{\lambda/2}{N_{ts}} \sum_{i=1}^{N_{ts}} \mathcal{R}\left(\mathbf{d}^{(i)}\right),$$
$$\mathcal{R}\left(\mathbf{d}\right) = \|\mathcal{D}_{\Theta}\left(\mathbf{d}\right) - \Psi\left(\mathcal{D}_{\Theta}\left(\mathbf{d}\right)\right)\|_2^2, \tag{2.4}$$

where N_{tr} and N_{ts} represent the number of training samples and testing samples, respectively. $\mathcal{L}(\cdot)$ and $\mathcal{R}(\cdot)$ are the MSE loss function and the regularization terms, respectively. Like (2.2), the formulation (2.4) is also suitable for different types of data acquisitions.

Since the DNN predictions are aimed to converge on low-frequency data, we enforce a soft constraint on the pass band of the predictions by $\mathcal{R}(\cdot)$. Compared with adding $\Psi(\cdot)$ to the predictions directly, this regularization guides the convergence of the DNN to smoother results, thus improving the stability of the training and reducing the high-frequency noise in the DNN predictions. Because the calculation of the regularization does not need the low-frequency data, $\mathcal{R}(\cdot)$ is applied to both the training inputs $\mathbf{d}_{HF}^{(i)}$ and the testing inputs $\mathbf{d}^{(i)}$. This technology can be viewed as a semi-supervised learning approach, where the smoothness of the training data and the testing data are enforced simultaneously during the training process. This approach is adopted in Jin et al. (2022b) for improving the quality of the predicted low-frequency electromagnetic signals.

2.3 Progressive Transfer Learning Method for Low Frequency Extrapolation

Previous works have demonstrated that deep learning approaches are superior to conventional BWE methods. However, the bottleneck caused by the dependency on training data influences the generalization ability of learning-based methods. The **progressive transfer learning (PTL)** (Hu et al., 2020, 2021; Jin et al., 2022a,b) is proposed for extrapolating low-frequency information in FWI with limited training data.

Under some circumstances, it is feasible to reconstruct the low-frequency components from the high-frequency data by the bandwidth extension (BWE) algorithm since they are implicitly connected by the subsurface geological structures and geophysical properties (Hu et al., 2021).

The conventional trace-by-trace BWE algorithms assume that the subsurface structures could be modeled as sparse reflection layers (Zhang and Castagna, 2011), a reasonable approximation, especially for post-stack subsurface images. When the subsurface features become more geologically complex, and the object of study is pre-migration data instead of seismic images, the learning-based methods are better alternatives because the sparsity assumption might not be justified, and the lateral continuity needs to be taken into account.

The recently well-developed learning-based approaches for low-frequency data extrapolation are much more flexible compared to conventional BWE algorithms. The most significant benefit is the incorporation of the implicit prior knowledge. The hidden relationship between the complicated subsurface formation and the seismic signals can be inferred by feeding synthetic or historical data to the DNN. Experiments show that the performance can be attractive if the training data and the field observations are highly similar (Jin et al., 2018; Ovcharenko et al., 2019b). However, this similarity is difficult to maintain when the DNN needs to be applied to the field data, which is from new and unknown subsurface structures. Due to the overfitting issue, the performance of the supervised learning can significantly drop (Sun and Demanet, 2015). Mitigating the overfitting issue, i.e., improving the generalization ability, is an important topic in supervised learning. A typical solution is to improve the amount of the training data. Note that the true low-frequency data cannot be acquired by field surveys. Therefore, the most typical way to increase the training data amount is to run the forward simulation. Each synthetic model can be used to generate a series of shot-gather data. However, how to customize the synthetic models used for generating training data is still a great challenge. Although the aforementioned transfer-style approaches can be used for generating synthetic models in a specific distribution (Kazei et al., 2020; Sun et al., 2021), we still lack the knowledge of the distribution of the true subsurface structures that we want to detect. PTL is proposed for solving this issue by leveraging the combination of deep learning and conventional physics-based inversion. Since the FWI can be used for revealing the subsurface structures from the observations, we exploit its model-reconstruction ability to help a velocity model to be updated progressively, like how the model is updated in a normal FWI workflow. The progressively updated model can be used for synthesizing training data, which will approach the observation in the meantime. Experiments show that applying the PTL algorithm to different supervised-learning DNNs can improve the performance in most cases (Hu et al., 2021; Jin et al., 2022a,b).

PTL is a transfer learning workflow combining physics-driven FWI and data-driven deep learning together. Since the FWI is embedded in the learning workflow to iteratively update the velocity model, the training dataset generated from the iteratively updating velocity model evolves as the learning process proceeds. Recently, this deep-learning-based frequency extrapolation was adapted to solve inverse scattering problem (Lin et al., 2021) and ground penetrating radar FWI (Jin et al., 2021, 2022b). The whole workflow of PTL is shown in Fig. 2.5. The workflow is composed of three important parts: (1) forward simulation engine, (2) network engine, and (3) FWI engine. Given the band-limited observations, the algorithm consists of the following steps: (1) Run the initialization and provide an initial velocity model based on the observed band-limited data or prior knowledge. In the simplest case, the initialization can be provided by an arbitrary model. In the following parts, more initialization methods will be discussed. (2) Feed the current velocity model to the forward wave-equation-based modeling simulation engine. The engine generates the wide-band synthetic dataset, including high-frequency and low-frequency components, which serve as the training dataset for the subsequent DNN training process. (3) Generate the training labels from the synthetic data and launch the DNN training process. (4) After the training converges, feed the high-frequency components of the recorded data into the trained network engine to predict the corresponding

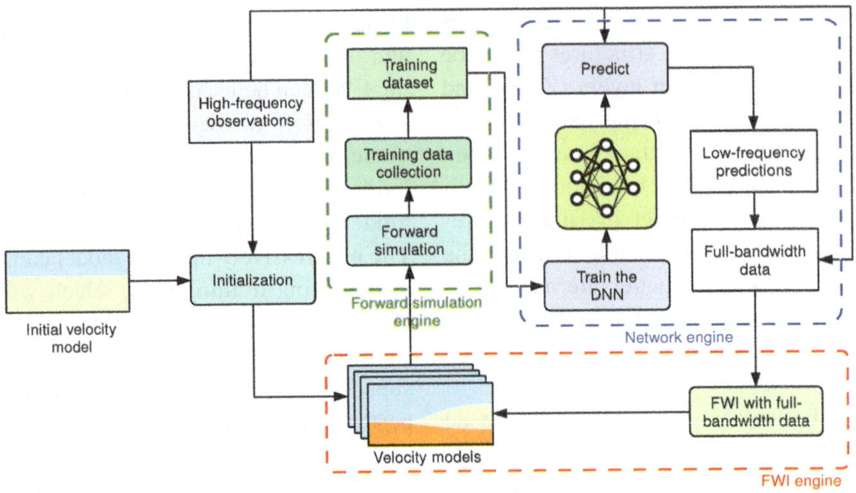

Fig. 2.5 The general workflow of PTL

low-frequency components. (5) Use the low-frequency predictions and the high-frequency observations to solve the inversion with the multi-frequency FWI engine. The FWI results may not be good enough in the first loop. However, it can be improved by repeating the algorithm. The updated velocity model is sent back to step (2) to start another round of the learning process. The algorithm repeats until both the training dataset and the FWI velocity model converge. After each loop, more subsurface information in the FWI results will be corrected. The updated training data will be better than the previous training data, thus ensuring better FWI solutions and extrapolated low-frequency data.

The proposed PTL is a general and flexible framework that can be combined with any popular DNN structure and data processing methods. The aforementioned specific steps can be adjusted according to the user requirements, prior knowledge, and the details of testing cases. In the original PTL work (Hu et al., 2021), the PTL workflow is initiated by a manually selected velocity model, and the training data is not pre-processed. The efficiency and accuracy of the workflow are further improved in Jin et al. (2022a). In the following sections, we will discuss the technical details of the velocity model initialization, the FWI module, the DNN training, and some data pre- and post-processing procedures.

2.3.1 Velocity Model Initialization for PTL

Similar to the standard FWI that needs an initial velocity model to start the inversion, the PTL workflow requires an initial velocity model to synthesize an initial training dataset. In the original PTL workflow (Hu et al., 2021), the initial velocity model is manually selected. Manually selecting an initial velocity model is probably the most convenient way to start the PTL workflow. However, a randomly selected initial velocity model without any a priori information embedded can be risky because the convergence of the PTL can be very slow, or the algorithm simply fails to converge. In this section, we will introduce three methods for the initial training velocity model building by combining the deep learning approaches with the conventional low-frequency extrapolation algorithms. The conventional low-frequency extrapolation methods provide better initialization than a randomly selected velocity model, thus accelerating the convergence of the PTL loops.

2.3.1.1 Initialization with Bandwidth Extension

When the initial velocity model contains more accurate low-frequency information, FWI results will be improved because low-

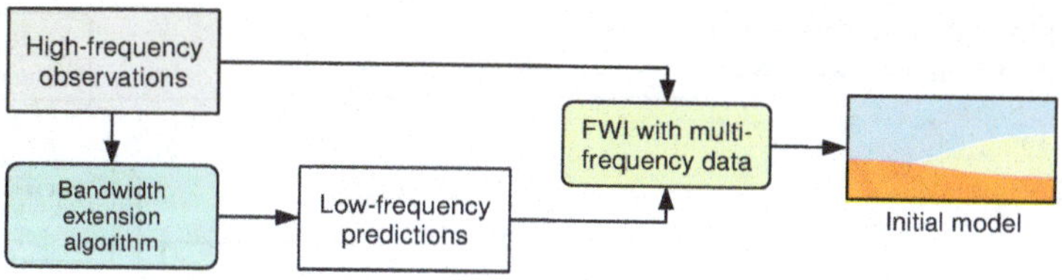

Fig. 2.6 Initialization with BWE algorithm

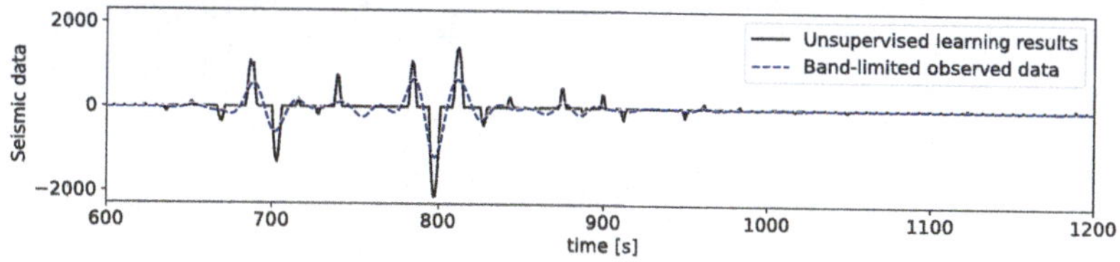

Fig. 2.7 Comparison between BWE results and band-limited observed data

frequency components provide correct kinematic information to the FWI engine for accurate positioning of the seismic events. BWE is a good candidate for providing such a warm start to FWI. In Fig. 2.6, we show the workflow of the BWE-based initialization. The algorithm starts with the band-limited observed data to implement the BWE. The approximate low-frequency solution obtained from the BWE, along with the observed high-frequency data, is sent to the FWI engine to reconstruct a velocity model that will serve as the initial velocity model of the next PTL iteration.

Let $\mathbf{d}$ and $\mathbf{d}_{\mathrm{HF}}$ denote the bandwidth-extended data and the observed high-frequency data, respectively. The BWE can be formulated as a sparsity-constrained inverse problem as follows,

$$\arg\min_{\mathbf{d}} \|\mathcal{F}\mathbf{d}_{\mathrm{HF}} - \Phi\mathcal{F}\mathbf{d}\|_2^2 + \lambda \|\mathbf{d}\|_1 . \tag{2.5}$$

where $\mathcal{F}$ and Φ represent the Fourier transform and a band-pass filter, respectively. The inverse problem is solved by a data-fitting regression in the frequency domain and an ℓ_1 regularization in the time domain. The band-pass filter is configured according to the reliable bandwidth of the observed data. There are several different algorithms available for solving this optimization, including the basis pursuit inversion (Zhang and Castagna, 2011; Chopra et al., 2006) and iterative shrinkage threshold algorithm (ISTA) (Pérez et al., 2013). In Fig. 2.7, we plot the BWE solution of a single seismic trace. The bandwidth-extended version of the trace could be viewed as a time series of reflectivity events.

2.3.1.2 Initialization with Learning-Based Bandwidth Extension

BWE is a good approach for providing an initial approximation of the low-frequency data. However, since BWE requires solving an inverse problem for each trace, the efficiency of the algorithm may not be suitable for processing most seismic datasets that usually consist of tens of millions or even billions of seismic traces. We propose a supervised learning method to improve the efficiency of the BWE algorithm.

Figure 2.8 depicts the workflow of the supervised-learning-based BWE. First, we select a small portion of the recorded band-limited seismic traces. Then, we perform the conventional BWE algorithm to obtain the low-frequency counterparts of these traces. These selected band-limited observations and the corresponding BWE results are the input and the label of DNN training, respectively. After the training converges, the DNN serves as a low-frequency data predictor for the remaining observed data. Benefiting from the feed-forward structure and the parallel computing design, the DNN can predict low-frequency data with high efficiency. In Fig. 2.9, we show an example of the learning-based BWE results. The full-bandwidth data is predicted by a 1D U-Net. The prediction is not very accurate but good enough for initializing PTL loops.

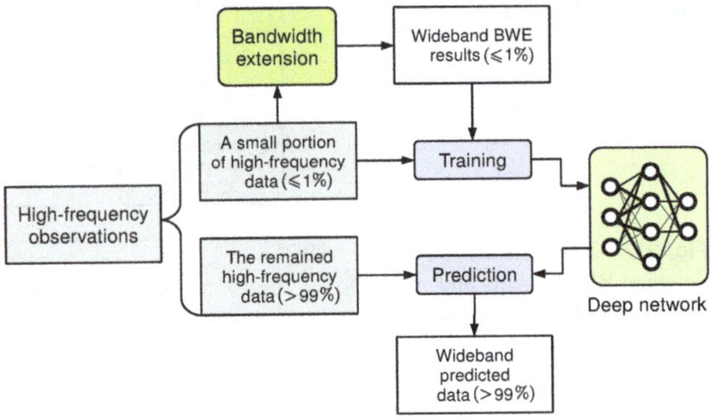

Fig. 2.8 The workflow of the learning-based BWE

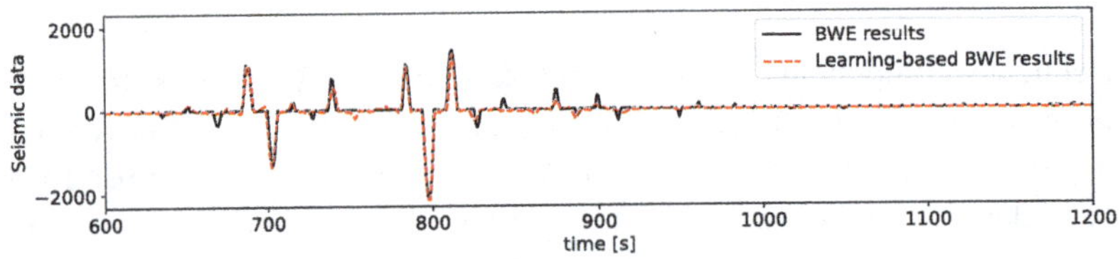

Fig. 2.9 Comparison of full-bandwidth predictions between conventional BWE result and BWE U-Net prediction

2.3.1.3 Initialization with Unsupervised Learning

Recently, the unsupervised learning methods (Mache et al., 2021; Gao et al., 2021) have been exploited to implement BWE to relax the assumptions in the conventional convolution model. Instead of using the results of conventional BWE algorithms, the DNN is used to reconstruct the full-bandwidth data from the band-limited observations directly. Given the DNN $\mathcal{D}_{\Theta_b}(\cdot)$ with the network parameters Θ_b, one example of the unsupervised learning is shown in the following formulation,

$$\arg\min_{\Theta_b} \frac{1}{N_{ts}} \sum_{i=1}^{N_{ts}} \left(\left\| \Phi\left(\mathcal{D}_{\Theta_b}\left(\mathbf{d}^{(i)}\right)\right) - \Phi\left(\mathbf{d}^{(i)}\right) \right\|_2^2 + \lambda \left| \mathcal{D}_{\Theta_b}\left(\mathbf{d}^{(i)}\right) \right| \right), \tag{2.6}$$

where N_{ts} and $\Phi(\cdot)$ are the amount of the training data and the band-pass filter, respectively. The problem could be viewed as a combination of the auto-encoder training and the deconvolution. To further improve the efficiency, we combine the unsupervised learning algorithm and the supervised learning strategy in this application. We select a small portion of the observed data as the training data to go through the unsupervised learning network and then let the trained network predict the full-bandwidth data for all the remaining data. In Fig. 2.10, we compare the observed band-limited data and the unsupervised learning results that are predicted by the same 1D U-Net used in the supervised learning network. Compared to the conventional sparse-inversion-based BWE, the DNN prediction is observed to be slightly less sensitive to small events.

2.3.1.4 Initialization with Beat Tone Data

When performing phase-only FWI, the beat tone data extraction (Hu, 2014, 2015) is also a feasible approach for initialization. Given the two frequency components in the observed data $\mathbf{d}_{f1}, \mathbf{d}_{f2}$, the beat tone approximation to the low-frequency data phase can be formulated as

$$\phi(\mathbf{d}_{f_{BT}}) = \phi(\mathbf{d}_{f1}) - \phi(\mathbf{d}_{f2}). \tag{2.7}$$

where $\phi(\cdot)$ is the phase extractor. The frequency of the beat tone data is the difference between the frequencies of the two high-frequency components $\mathbf{d}_{f1}$, $\mathbf{d}_{f2}$. With the extracted beat tone phase, we could perform the multi-frequency FWI in Fig. 2.6. Although the beat tone approximation to low-frequency data loses accuracy in complex geological environments, this method could be a good alternative to the

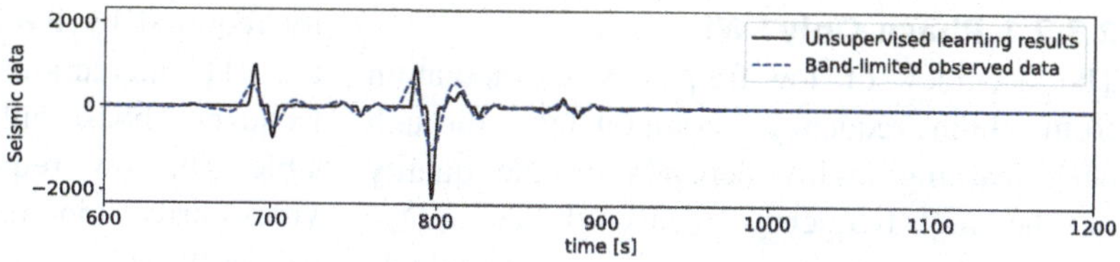

Fig. 2.10 Comparison between unsupervised learning results and band-limited observed data

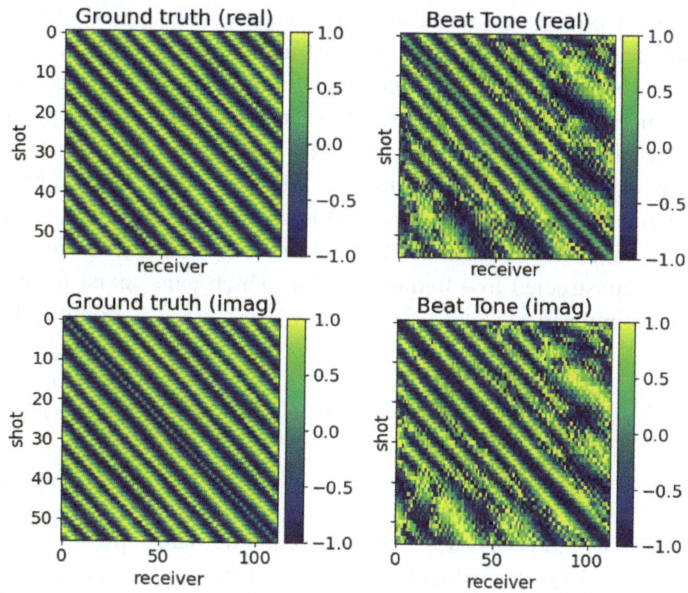

Fig. 2.11 Comparison of the 3 Hz phase data. The real part and the imaginary part of the phase are shown in the first row and the second row respectively. The true data and the beat tone data are shown in the first column and second column respectively

algorithm to launch the PTL workflow because of the high efficiency of beat tone phase extraction and the iterative self-correction feature of the PTL. In Fig. 2.11, we show an example of the beat tone extraction from a synthetic dataset generated on a scaled BP2004 model. The 3 Hz beat tone data is calculated from the 15 Hz component and the 18 Hz component.

2.3.2 Special Practices of FWI for Efficiency and Robustness Enhancements

There are many different types of FWI algorithms to accommodate seismic data with different properties. For example, depending upon the types of loss function, FWI can be classified into ℓ_2-norm FWI (Virieux and Operto, 2009; Pan et al., 2016), cross-correlation FWI (Liu et al., 2017; Dong et al., 2020), envelope FWI (Chi et al., 2014), phase-only FWI (Bednar et al., 2007; Maharramov et al., 2017; Sheng et al., 2020), and so on. From the perspective of optimization algorithm, there are conjugate gradient FWI (Pan et al., 2017; He and Han, 2018) , Gauss-Newton FWI (Pan et al., 2016; Chen and Tran, 2021), full-Newton FWI (Pan et al., 2016; Anagaw and Sacchi, 2012), truncated Newton FWI (Métivier et al., 2013, 2015), etc. In the workflow of the progressive transfer learning for low-frequency extrapolation, we employ three different FWI engines in the seismic FWI:

1. The first FWI engine is a phase-only low-frequency FWI that is used to invert the extrapolated low-frequency components.
2. The second FWI engine is a truncated high-frequency conjugate-gradient FWI to obtain a velocity model for training data generation.
3. The third FWI engine is a standard high-frequency FWI engine to invert the full waveform information contained in the measured high-frequency seismic data. This engine delivers a final velocity model for subsurface interpretation.

2.3.2.1 Phase-Only FWI

The accuracy of low-frequency extrapolation from high-frequency components through deep learning highly depends on the quality of the high-frequency measured data. Due to the unknown source wavelet, unavailable density profile, and uncertain subsurface seismic attenuation properties, the reconstructed phase information of the low-frequency components is more reliable than the amplitude information. The phase-only FWI can avoid the leakage of the wrong amplitude information into the FWI engine. During the phase-only FWI, a phase-only loss function is recommended for the inversion of reconstructed low-frequency data, which can be formulated as

$$C(\mathbf{v})=\sum_{f=1}^{N_f}\sum_{s=1}^{N_s}\sum_{r=1}^{N_r}\left\|\phi\left(\mathbf{d}_{r,s,f}^{(syn)}(\mathbf{v})\right)-\phi\left(\mathbf{d}_{r,s,f}^{(obs)}\right)\right\|_2^2, \tag{2.8}$$

where $\mathbf{d}_{r,s,f}^{(syn)}(\mathbf{v})$ and $\mathbf{d}_{r,s,f}^{(obs)}$ are the synthetic data generated on the velocity model $\mathbf{v}$ and the observed data, respectively. $\phi(\cdot)$ is a phase extractor.

2.3.2.2 Truncated High-Frequency FWI

After the first step of phase-only FWI, a low-resolution velocity model that contains valuable subsurface low-wavenumber information is obtained. Under the framework of PTL, the next step is to incorporate the high-wavenumber components by inverting the measured high-frequency seismic data. Due to the high computational cost, even with the GPU acceleration, high-frequency FWI is still time-consuming. The standard high-frequency FWI inverting observed data in Fig. 2.3 is replaced by a truncated FWI to improve the efficiency of PTL. The truncated FWI can be regarded as a standard high-frequency FWI stopped after one iteration.

The resulting velocity model after these two steps of FWI is the so-called training velocity model that is used for generating training data. The purpose of developing such a training velocity model is to establish a wavenumber spectrum pattern similar to the true subsurface geological environment. In production FWI, the data quality required by FWI is very high. However, in the PTL interactions, we only perform the low-frequency phase-only FWI as precisely as possible. The low-frequency phase-only FWI provides correct kinematic information that is critical for the subsurface low-wavenumber structure reconstruction to avoid cycle-skipping. In comparison, the high-frequency FWI generates fine-scale geological details in the resulting training velocity model. These high-resolution features do not significantly interfere with the important reconstructed low-frequency information. In other words, these high-resolution features obtained after a single iteration of high-frequency FWI only exist in the training velocity model. The corresponding high-frequency data generated on this training velocity model only serves as the training data for the deep learning-based low-frequency data extrapolation. They will not be inverted in the final production FWI. With this strategy, we decouple the high-frequency FWI for training velocity model generation and the high-frequency FWI for final product delivery, which allows us to maximize the efficiency of PTL.

We use a scaled BP2004 model to validate our method. The true velocity model is displayed in Fig. 2.12. In Fig. 2.13, we show one example of the truncated FWI. The experiment results show that the low-wavenumber features in the solved model mainly depend on low-frequency FWI. Compared to the FWI results with true low-frequency data, most subsurface structures are successfully retrieved by the FWI with learning-based BWE data. The results are not perfect. The final training velocity model generated by truncated FWI is strongly affected by migration-like swings and oscillating artifacts. The salt body on the left side is not corrected and inverted. However, we emphasize again that these strong artifacts only exist in the training velocity model, and they will not prevent us from establishing the correct mapping between the low-frequency and high-frequency components in the data, as will be validated in the following experiment. Since a high-frequency FWI is significantly more expensive than its low-frequency counterpart, this approach can drastically reduce computational costs. For example, the computational cost of

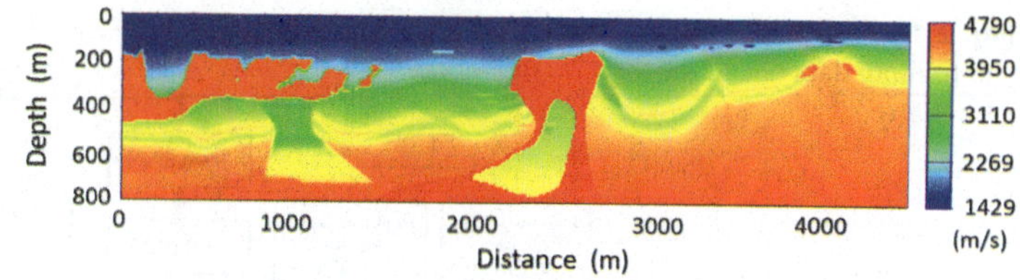

Fig. 2.12 The true scaled BP2004 model. 56 shots and 112 receivers are deployed on the surface of the model

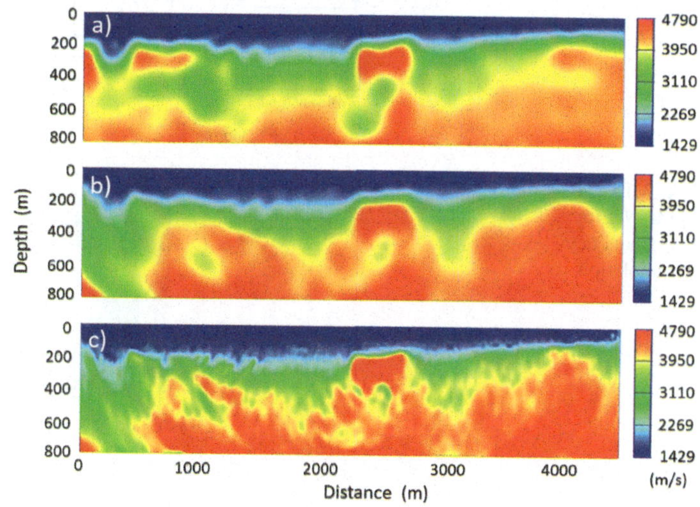

Fig. 2.13 Comparison of FWI results: (**a**) FWI with true low-frequency data (3 ∼ 9 Hz), (**b**) FWI with learning-based BWE results (**c**) FWI with learning-based BWE results followed by truncated high-frequency FWI with observed data (10 ∼ 30 Hz)

inverting a 10 Hz dataset using a conjugate gradient-based FWI engine is approximately 16 times that of inverting a 5 Hz dataset.

2.3.3 Low-Frequency Extrapolation with DNN

The detailed workflow of the PTL loops is shown in Fig. 2.14. Each loop consists of a forward simulation module, a DNN training module, and an FWI module. The PTL is launched with an initial velocity model constructed using one of the approaches described in Sect. 2.3.1. The input (high-frequency data) and the label (low-frequency data) for training the DNN are generated by a forward full wave propagation simulator. After collecting the training dataset, we train the DNN and use the trained DNN to predict the absent low-frequency data from the corresponding high-frequency observations. The low-frequency predictions, along with the high-frequency observations, are then input into an FWI engine to construct a training velocity model for the next PTL loop.

After the first PTL iteration, the predicted low-frequency data are usually not accurate to deliver a final FWI velocity model because the training velocity model used in the first PTL iteration can be very different from the true subsurface velocity model. Consequently, the training data generated on this training velocity model might not reside in the same sampling space as the measured high-frequency data and the low-frequency data to be reconstructed. This non-representative nature of the training data in the first PTL iteration often results in biased network prediction. A straightforward strategy for overcoming this network generalization problem is to build many training velocity models. The corresponding training datasets are designed to cover the sampling space of the measured data. Unfortunately, this strategy is not effective in practice because it is inherently impossible to establish a gigantic training velocity model library covering the full range of the global geological and geophysical environments.

Under the framework of PTL, a different strategy is proposed to overcome this challenge. Instead of building a large quantity of training velocity models to train the low-frequency extrapo-

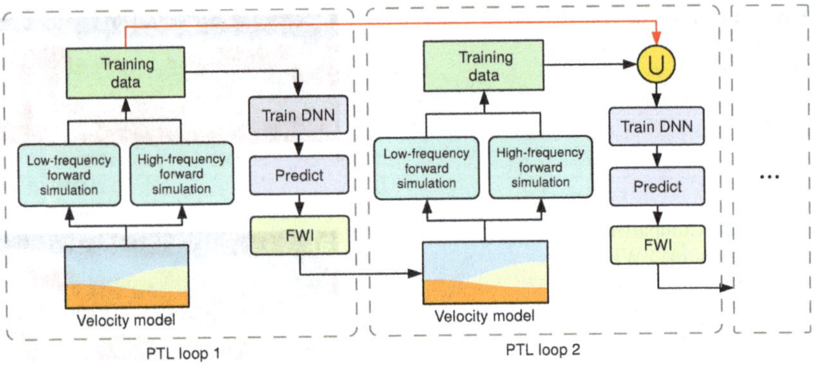

Fig. 2.14 The PTL loop and training data collection

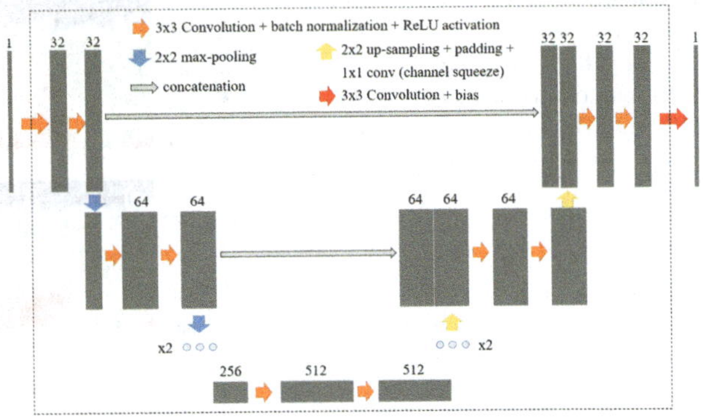

Fig. 2.15 The structure of 2D U-Net for low-frequency extrapolation

lation network in parallel, in each PTL iteration, only one training velocity model is used for the network training. However, this training velocity model evolves as the PTL iteration proceeds. In the first PTL iteration, a set of low-frequency data is predicted by the network. Although this predicted low-frequency data is not accurate, it contains some subsurface low-wavenumber information, which is valuable for the subsequent FWI process to build a better training velocity model for the next PTL iteration. On the other hand, a better training velocity model can improve the accuracy of the network prediction of the low-frequency data. In this manner, after each PTL iteration, the low-frequency data predicted by the DNN module helps the FWI module build a better training velocity model by retrieving more subsurface information contained in the measured data. On the other hand, the resulting improved training velocity model provides a more representative training dataset to the DNN module to predict more accurate low-frequency data in the next PTL iteration. Eventually, the evolving training velocity model converges to the final FWI velocity model, and the corresponding high-frequency training dataset converges to the measured dataset while the predicted low-frequency components achieve high accuracy. The cycle-skipping phenomenon is successfully suppressed.

In practice, instead of replacing the previous training data with the updated training data, all the history training datasets can be preserved to enlarge the training data library, which improves the robustness of the algorithm. The red line in Fig. 2.14 emphasizes this accumulating training data strategy.

2.3.3.1 U-Net Structure

The DNN structure in PTL is designed to be flexible and easily adapted to different data arrangements. In this section, we introduce the 2D U-Net architecture employed in our work. The DNN structure is illustrated in Fig. 2.15. The input and predicted data are collected in the time-spatial domain. The DNN receives the high-frequency shot-gather as the input and produces the corre-

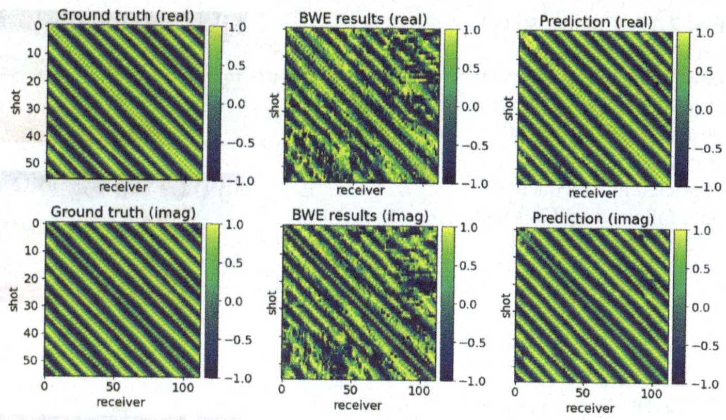

Fig. 2.16 Comparison of the 3 Hz phase data. The real part and the imaginary part of the phase are shown in the first row and the second row respectively. The true data, the learning-based BWE results, and the PTL results are shown from left to right side respectively

sponding low-frequency shot-gather as the output. The U-Net is constructed by a four-level encoder, a four-level decoder, and an intermediate layer. The intermediate layer joins the encoder and the decoder. At each level, the feature maps of the encoder are connected to those of the decoder by a skip connection. These skip connections are critical components of the U-Net because they transmit the various levels of semantic information to the decoder and preserve the image details in the final predictions. The output layer is an affine convolutional mapping without activation. At each level, the input feature maps are convolved twice with the 3×3 filters. The first convolution doubles the channel size of the feature map. Each convolution is followed by a batch normalization and a ReLU activation. The channel numbers of the feature maps at different levels are 32, 64, 128, 256, respectively. Two consecutive convolutional layers in the intermediate layer further double the channel size from 256 to 512. The adjacent levels are connected by downsampling (or upsampling) operations.

2.3.3.2 Testing Results of BP2004 Model

The basic idea of PTL is to improve the training dataset by running the simulation and the DNN training alternately. We start from the model shown in Fig. 2.12, where transmission waves dominate the survey. Since transmission waves convey the most important information, the PTL can converge to promising results quickly.

The effectiveness of the PTL is validated by a numerical experiment performed on a scaled BP2004 model. The true velocity model is shown in Fig. 2.12. The DNN predicts the absent low-frequency data (3 Hz $\sim$ 9 Hz) from the measured high-frequency data (10 Hz $\sim$ 30 Hz). The initial training velocity model is generated by the learning-based BWE. This initial training velocity model is further improved gradually in the subsequent PTL iterations. Eventually, the training velocity model is able to provide an adequately representative training dataset to the DNN module. The final low-frequency prediction results (3 Hz) are displayed in Fig. 2.16. In Fig. 2.17, we compare the FWI results obtained by inverting the true low-frequency data and the low-frequency predictions from the PTL. In this numerical experiment, it only takes one loop for the PTL to converge. The experiment results demonstrate that the accuracy of the PTL-predicted low-frequency data is superior to that produced by the BWE approach. The discontinuity in the BWE results disappears in the PTL results. Compared the corresponding FWI velocity model with the one shown in Fig. 2.13, the oscillation and artifacts in Fig. 2.17b are effectively suppressed, and the salt body on the left side is successfully reconstructed. With the low-frequency data predicted by PTL, the final velocity model produced by the FWI is almost identical to the velocity model reconstructed using the true low-frequency data.

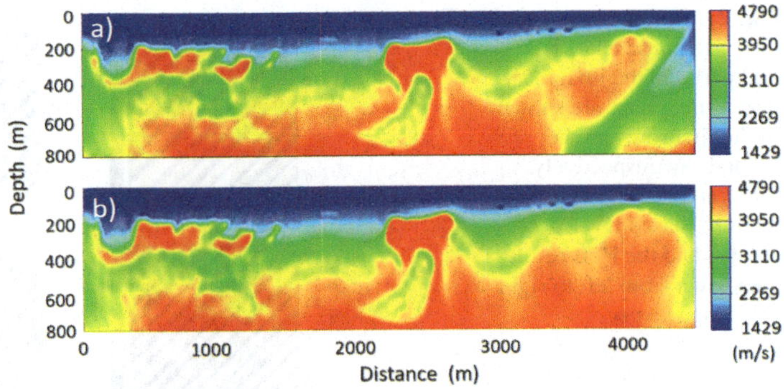

Fig. 2.17 Comparison of (**a**) FWI with true low-frequency data (3 ~ 9 Hz) followed by FWI with observed data (10 ~ 30 Hz), (**b**) FWI with predicted low-frequency data followed by FWI with true observed data

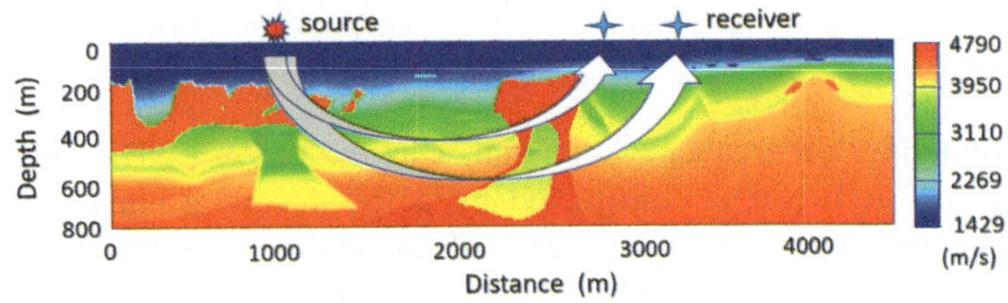

(a)

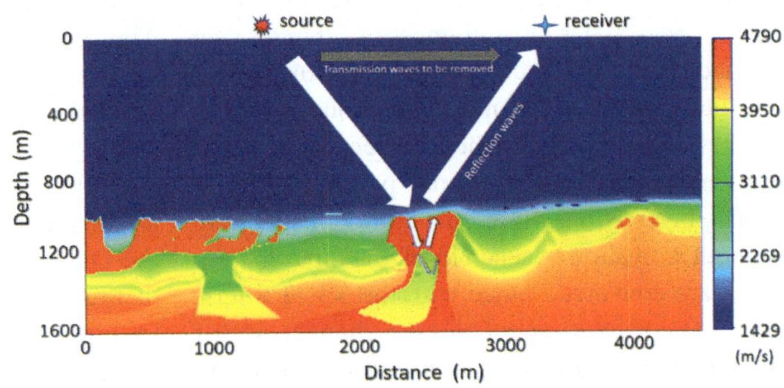

(b)

Fig. 2.18 Comparison of testing cases based on the scaled BP2004 model. (**a**) Dominated by transmission wave propagation patterns; (**b**) Dominated by reflection wave propagation patterns

2.3.4 PTL in a Reflection Dominant Case

In the scaled BP2004 velocity model shown in Fig. 2.12, the seismic energy in observations is dominated by transmission waves. In this section, we test the PTL in a more challenging FWI task. Figure 2.18 compares the transmission dominant case and the reflection dominant case. In transmission case, the dominant early arrivals convey a wealth of subsurface geological and geophysical data. However, the penetration depth of transmission waves is determined by the offset (the distance between a source and a receiver) and the subsurface condition. For example, the gradient of subsurface velocity plays an important role. If there is a high-velocity layer above a low-velocity layer, the transmission seismic wave is unable to travel deeper. Due to these limitations of transmission seismic data, reflection seismic has to be exploited to reveal the deep region structure and properties. To conduct a numerical experiment on pure reflection seismic data, in Fig. 2.18b, we add a deep water layer on the top of the scaled BP2004 model, and thus, the transmission seismic energy propagating through the water layer can be easily separated from the late arriving reflection seismic events.

Reflection seismic data are usually more challenging to handle in an FWI process because the inverse problem is more nonlinear in comparison with transmission seismic FWI. When strong reflectors, such as salt domes, are present in the subsurface structures, the reconstruction of the structure and properties beneath the strong reflectors becomes even more challenging, as the majority of the wave energy is reflected and cannot penetrate the strong reflectors. This problem complicates the learning-based low-frequency extrapolation because the sub-salt geological structures are concealed in those weak late-arrival signals. The data pretreatment is critical to amplifying the weak late arrivals, thus improving the performance of the PTL in reflection-dominant cases.

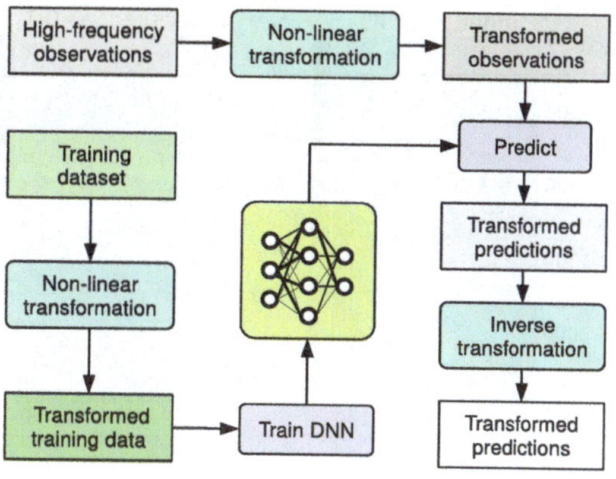

Fig. 2.19 Training and testing with the nonlinear transformation

2.3.4.1 Pre- and Post- Processing

The input and label of the DNN can be enhanced by a revertible nonlinear transformation, which is designed to boost the late-arrival energy. Instead of using the less flexible automatic gain control (AGC) (Ovcharenko et al., 2020) or geometrical spreading correction (Schäfer et al., 2012), a simple logarithmic transformation is proposed for this task. Denoting the time-domain data and the transformed data as $\mathbf{x}$ and $\mathbf{y}$ respectively, the forward and inverse transformations $\mathcal{T}(\cdot)$ could be formulated as

$$\mathcal{T}(\mathbf{x}) = \text{sign}(\mathbf{x}) \log_{10}(|a\mathbf{x}| + 1), \quad (2.9)$$

$$\mathcal{T}^{-1}(\mathbf{y}) = a^{-1}\text{sign}(\mathbf{y}) \left(10^{|\mathbf{y}|} - 1\right), \quad (2.10)$$

where the parameter a is the coefficient used for adjusting the level of amplitude boosting of the late arrivals, the proposed transformation reweighs the importance of the seismic events with different amplitudes. The weak reflection events are more amplified compared to strong events. Consequently, the contributions from the seismic events with largely varying amplitudes become more balanced, so they can be easily digested by the DNN training process without introducing biases that severely affect the network prediction.

In Fig. 2.19, we illustrate the training and testing with the nonlinear transformation. Before training the DNN, both the inputs and labels in the training dataset are pretreated by the logarithmic transformation. During the testing phase, the input data is boosted before being fed into the DNN. The DNN prediction is consequently transformed back into the original domain with the inverse transformation $\mathcal{T}^{-1}(\cdot)$.

Figure 2.20a, b show an example of the nonlinear transformation results. The amplitude of the first reflection event in the original data is approximately 10 $\sim$ 100 times that of the late arrivals, posing great difficulties in predicting the deep reflections. By adjusting the value of a, the late-arrival energy can be promoted to different levels. In both the training and testing examples, we employ the identical configuration for the pre- and post-processing.

2.3.4.2 Testing Results of Deep Marine BP2004 Model

The improved PTL is validated by the reflection dominant case, where the data is predicted in the time domain. In the initialization, we use the learning-based BWE results and a 1D U-Net to provide a warm start. During the FWI, we use the truncated FWI to accelerate the construction of the training velocity model. When training the DNN, we use the nonlinear transformation to balance the data amplitudes. The observed data is generated by a 15 Hz Ricker wavelet and the deep marine velocity model shown in Fig. 2.12. To evaluate the performance of the PTL, a ground truth dataset is synthesized by using a 3 Hz Ricker source wavelets. To ensure the blindness about

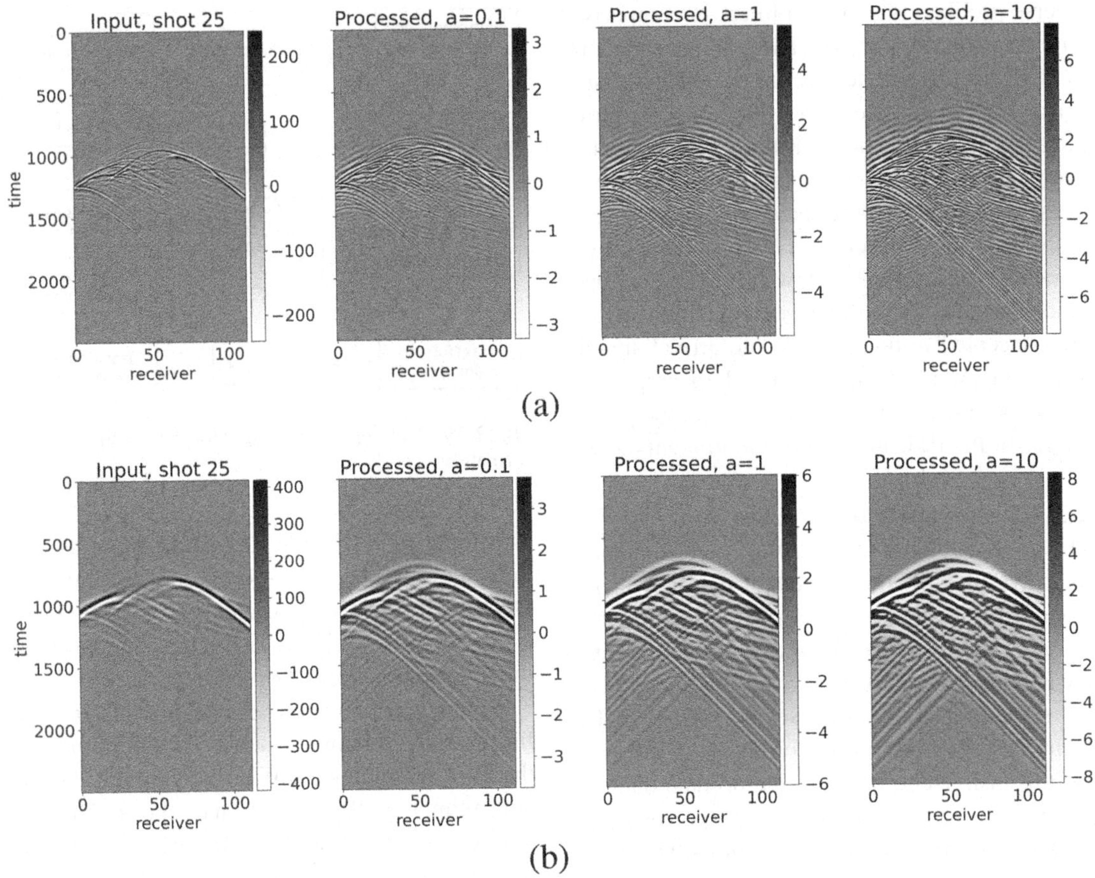

Fig. 2.20 An example of nonlinear transformation on reflection dominated data: (**a**) input data (**b**) 6 Hz label data. Results from the left to the right are the raw data, the transformed data with $a = 0.1$, $a = 1$, and $a = 10$ respectively

the low-frequency information, we apply a high-pass Butterworth filter to remove the frequency components lower than 10 Hz from the observations. With these configurations, a 2D U-Net illustrated in Fig. 2.15 is trained under the PTL framework to predict the $3 \sim 9$ Hz data from the observed $10 \sim 30$ Hz data. The convergence of the DNN takes 80 epochs and about 1 or 2 hours, which is faster than the FWI process.

A shot-gather of the low-frequency extrapolation results is illustrated in Fig. 2.21. To visualize the data clearly, we perform the nonlinear transformation with $a = 1.0$. As a comparison, the ground truth of the 6 Hz data is shown in Fig. 2.21a, along with, we compare the extrapolated low-frequency data with and without the nonlinear transformation shown in Fig. 2.21b, c. Several traces are extracted from the shot-gather and compared in Fig. 2.22 to reveal more details. The efficacy of the nonlinear transformation is validated in the experiment results. Without the nonlinear transformation, only the early arrival events are reconstructed. In comparison, the quality of the weak late arrivals is successfully predicted by the DNN with the nonlinear transformation.

In Fig. 2.23, the phase of the DNN-extrapolated 3 and 6 Hz components (the third column) are compared with the ground truth (the first column) and the BWE results (the second column). Because the BWE is a trace-wise inversion operator, discontinuity is observed between traces. Compared with the transmission data in Fig. 2.16, the BWE results are more severely

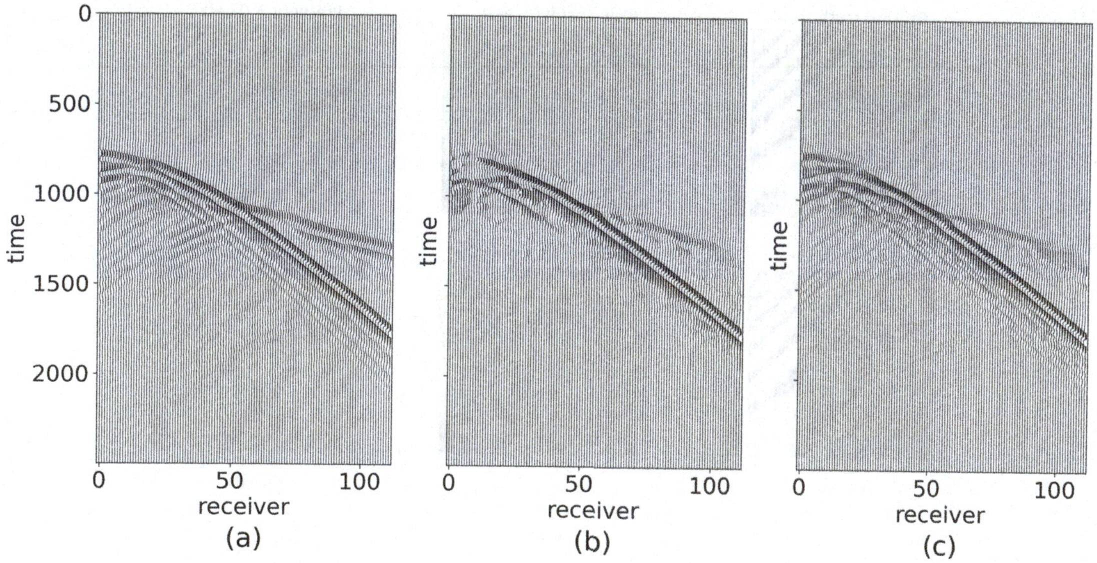

Fig. 2.21 An example of the low-frequency extrapolation: (**a**) true low-frequency data generated by a 6 Hz Ricker wavelet, (**b**) PTL predictions without the nonlinear transformation, (**c**) PTL predictions with the nonlinear transformation

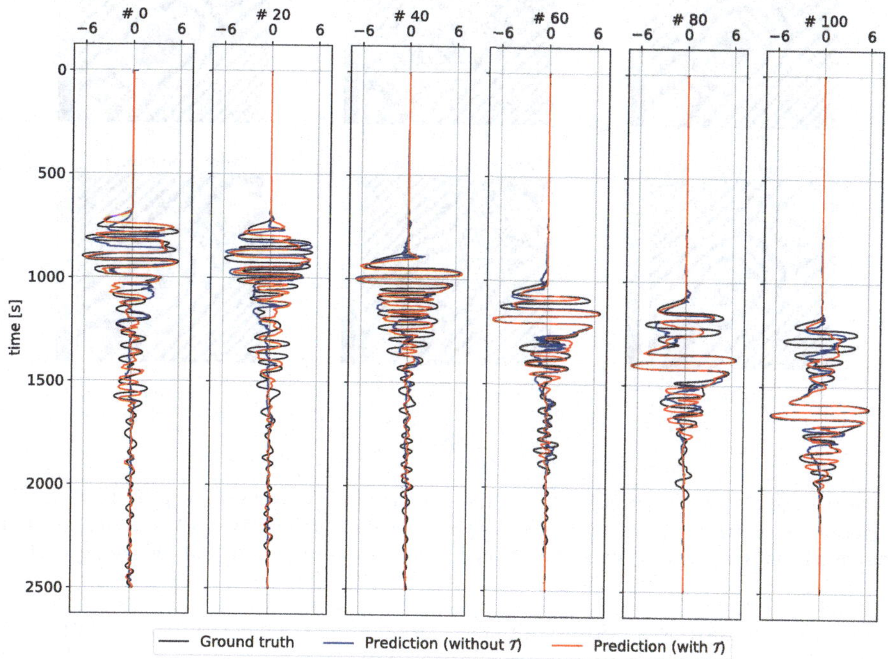

Fig. 2.22 Comparison of several traces in the logarithmic domain. The receiver number is shown on the top of the plot. $\mathcal{T}(\cdot)$ represents the nonlinear transformation

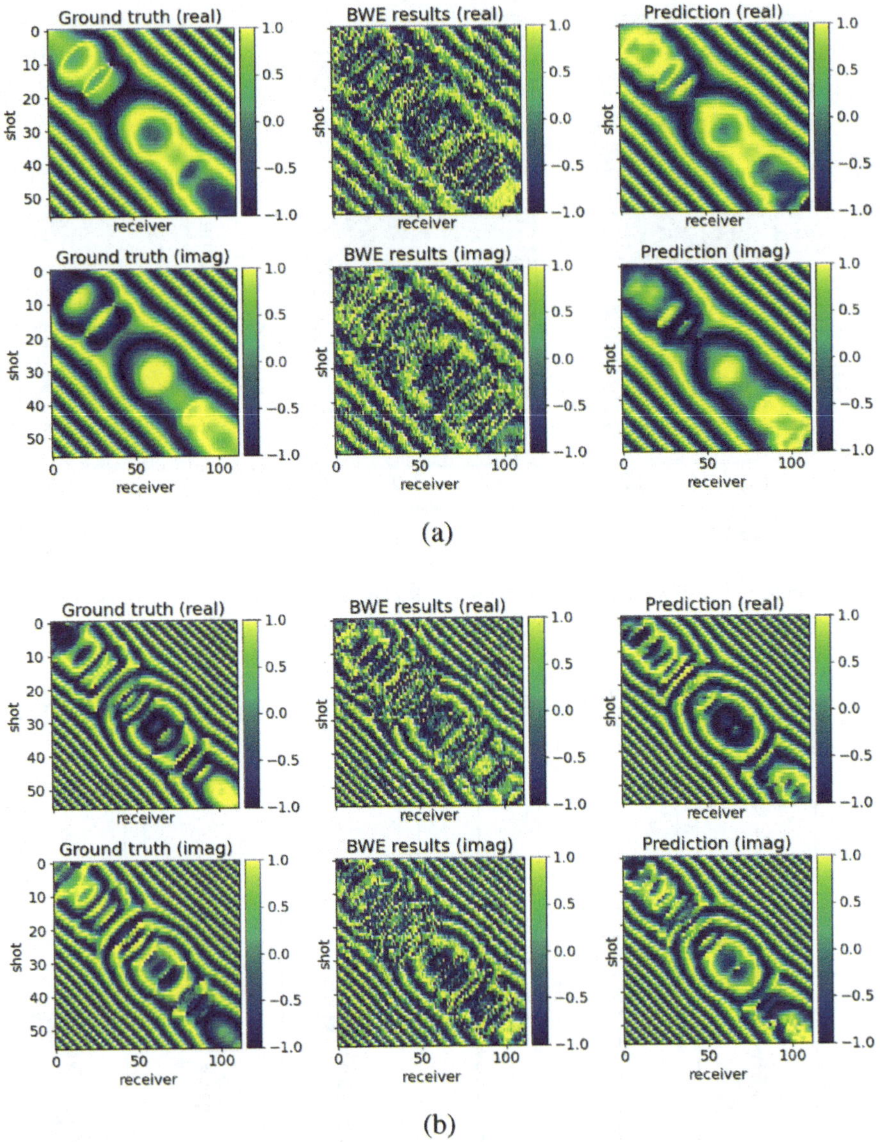

Fig. 2.23 Comparisons between the phase of the frequency components: (**a**) 3 Hz components, (**b**) 6 Hz components. The real part and the imaginary part are compared in the first row and the second row respectively. The true data, the learning-based BWE results, and the PTL results are shown from left to right side respectively

contaminated by the artifacts. Although the BWE results are imprecise, they only serve as the initialization of PTL. After running one or two PTL loops, the artifacts gradually disappear, and the PTL starts to converge. The final low-frequency data extrapolated by the PTL bear a high resemblance to the ground truth.

The accuracy of the low-frequency extrapolation is further validated by the FWI. In Fig. 2.24, we compare the performance of the standard PTL and the PTL improved by the BWE initialization, truncated FWI, and nonlinear transformation. We perform the low-frequency FWI with the PTL predictions followed by a series of full high-frequency FWI with the measured band-limited seismic data. The convergence of the improved PTL only requires two PTL loops. When comparing the standard PTL results in Fig. 2.24a to the

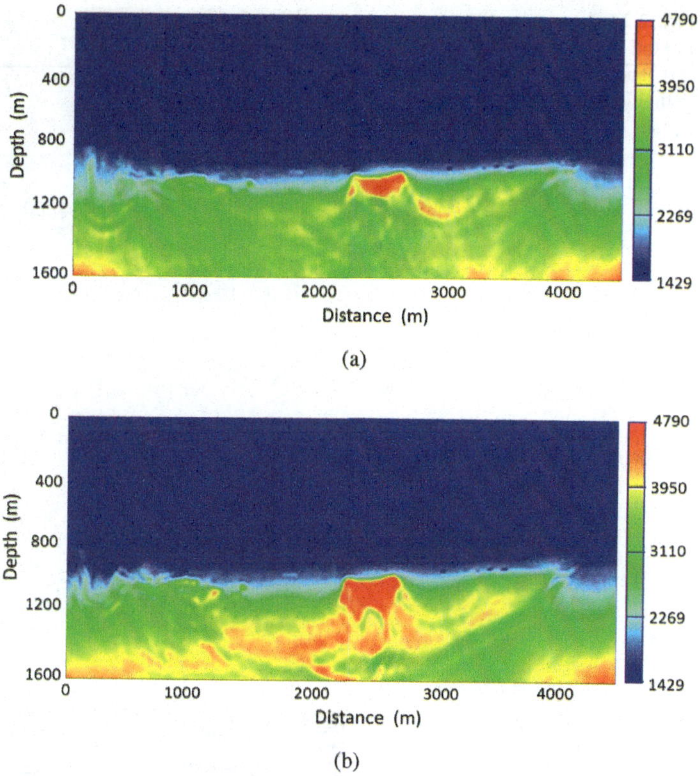

Fig. 2.24 The velocity model obtained by FWI with low-frequency predictions (3 ~ 9 Hz) followed by FWI with high-frequency observations (10 ~ 30 Hz): (**a**) predicted by a standard PTL workflow, (**b**) predicted with the improved PTL

true model in Fig. 2.18b, it is clear that only the top of the salt body is resolved, while all the subsalt structures are absent. The main reason for this shielding effect is that the first reflection event is so strong that it dominates the DNN prediction and the FWI inversion process. Figure 2.24b illustrates the reconstructed velocity model using the improved PTL. The performance is superior to the standard PTL. More deeper structural details are resolved, especially in the middle section of the velocity model. The subsurface structures near the edges of the velocity model are not successfully reconstructed, partially due to the poor illumination in those areas.

2.3.5 PTL in a Field Case

For our field data experiment, we leveraged the publicly available Poland 2D Vibroseis Line 001 land dataset to validate the proposed workflow. This dataset is comprised of 251 shot gathers, spaced 50 m apart, generated by a vibroseis source with a sweep frequency range of 8 to 95 Hz. Each shot gather contains data from 284 receivers, each spaced 25 m apart. The dataset is sampled at a 2 ms interval, covering a total recording time of 3 seconds. An illustration of a raw seismic gather from this dataset is presented in Fig. 2.25a, where ground roll noise is predominantly visible in the near offset region, alongside other interference noises at the periphery. To simplify the analysis, we directly applied a mute to the noisy sections, as depicted in Fig. 2.25b.

In this experiment, we adopt the BWE method for generating the initial model using FWI. During this process, apart from the bandwidth extended data, a background velocity model is required for the start of FWI . Conventional methods often adopt a randomly generated layered model. However, this could potentially lead to unstable convergence and local minimum. Therefore, we integrate the conventional tomographic approach to pick the velocities from the velocity panel in Fig. 2.26, and the resulting model

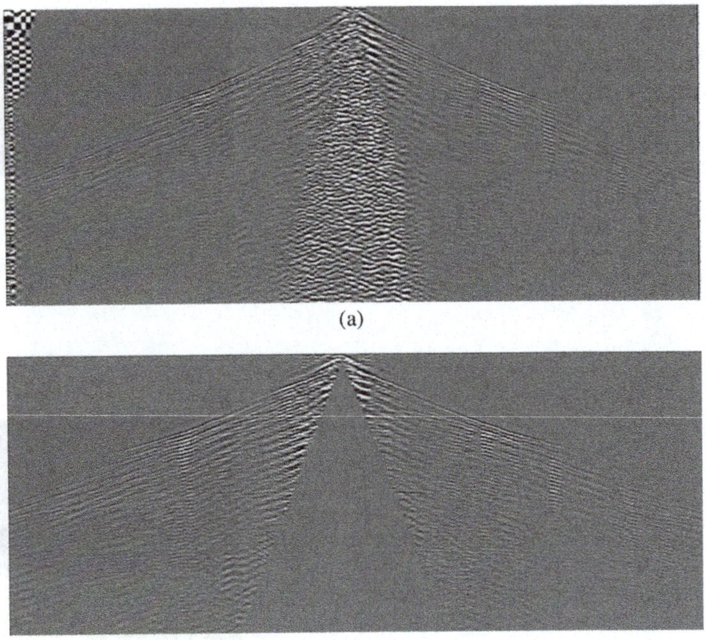

(a)

(b)

Fig. 2.25 Example of the Poland 2D vibroseis line 001 land dat: (**a**) Raw shot gather, (**b**) Denoised gather of (**a**)

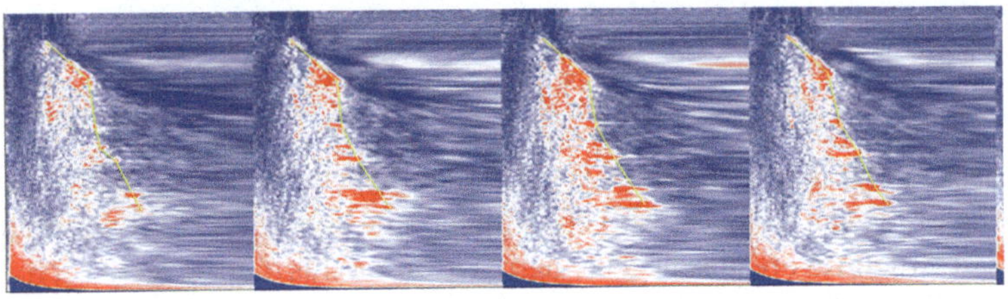

Fig. 2.26 Example velocity picking panel of the Poland dataset

is presented in Fig. 2.27a. We can observe that the picking result is rough, leaving space for improvement.

Next, the BWE method is deployed to the raw gather and extend its bandwidth in low-frequency region. As can be compared in Fig. 2.28a, b, the frequency components below 9 Hz has become richer in this process, whereas the dominant frequency is at around 12 Hz. The enriched data between 8–16 Hz then serves as the basis for the initial round of FWI, and the result is presented in Fig. 2.27b. To analyze the quality of the model, we present the imaging result of the dataset in Fig. 2.29. The comparison between Fig. 2.29a, b illustrates the improvement in image quality achieved through the BWE process, without significant degradation in specific regions.

Leveraging the improved model obtained from the BWE-enhanced FWI process, the experiment then embarked on a progressive transfer learning and subsequent FWI update. This methodology begins with the creation of two synthetic datasets. The input dataset for machine learning is synthesized with a 12 Hz Ricker wavelet, incorporating a low-cut filter to omit frequencies below 9 Hz, while the output simulation targets a dominant frequency of 5 Hz, effectively spanning the 0 to 9 Hz frequency range. We employ a four-stage 1D Unet for the learning task., focusing on minimizing the Mean Squared Error (MSE) between the network's prediction and the synthetic low-frequency target. The trained network is then applied to predict the field data's low-frequency content from 0 to 9 Hz, which is presented in

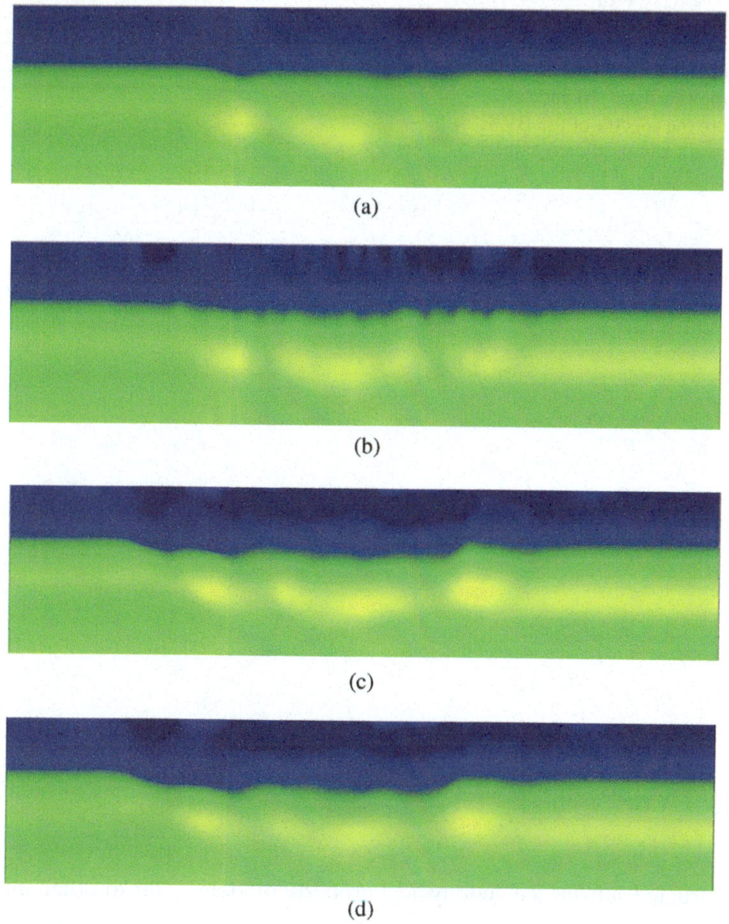

Fig. 2.27 Interval velocity models in depth. (**a**) is the interval velocity model by picking the NMO panel; (**b**) is the initial model by FWI with the BWE result, respectively; (**c**) is the updated model by FWI using the second training iteration result; (**d**) is the final update of (**c**) with 16–25 Hz high frequency components of the field data

Fig. 2.28c. Remarkably, this process yields an average improvement of above 10 dB in the quality of low-frequency components, with the dominant frequency of the prediction around 5 Hz, demonstrating a significant enhancement over both the initial raw input and the BWE results.

Leveraging the predicted low-frequency data (1–8 Hz) alongside the original data band (9–15 Hz), a new FWI iteration yields an updated interval velocity model in depth, as depicted in Fig. 2.27c, and the corresponding imaging result is presented in Fig. 2.27c. This integration of predicted low-frequency data leads to a notable improvement in stack image quality compared to prior iterations.

In the final phase, data spanning frequencies from 1 to 15 Hz, enriched with the main velocity structure determined through the preceding steps, is combined with data from the 16 to 25 Hz range extracted from the raw input. This integration aims to capture more detailed subsurface features, with the ultimate results showcased in Figs. 2.27d and 2.29d.

2.4 Summary

In this chapter, we demonstrate the workflow of progressive transfer learning (PTL), apply the method to the low-frequency extrapolation of the seismic data, and show the effectiveness of this data enhancement approach by solving full waveform inversion (FWI) in different cases. Due to the limitation of the data acquisition device, one of the main challenges in FWI is the lack of low-frequency components in most seismic datasets, which may lead FWI algorithms to local minima and result in strong artifacts in the recon-

Fig. 2.28 Frequency analysis for the seismic data. (**a**) The spectrum of the raw data; (**b**) and (**c**) are the spectra of the BWE result and predicted gather, respectively

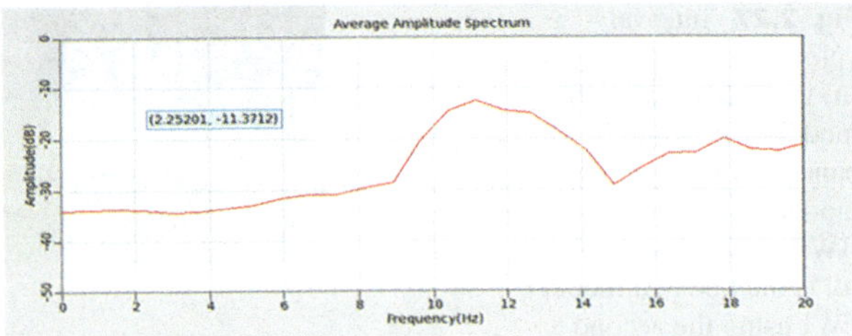

(a)

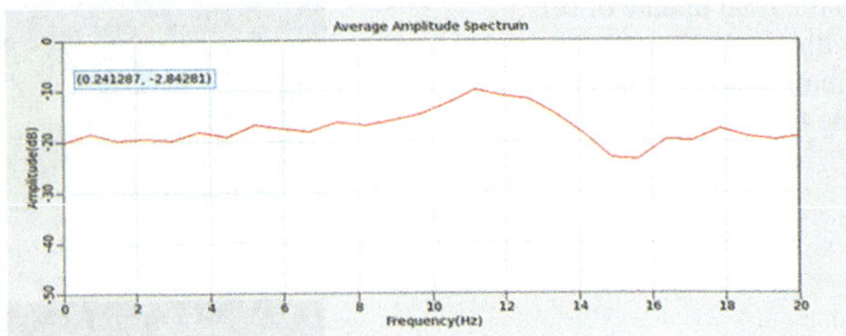

(b)

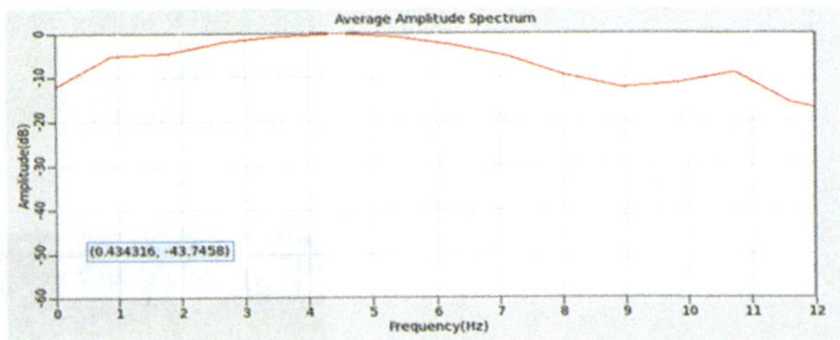

(c)

structed subsurface geophysical property models, a phenomenon called cycle-skipping. The proposed PTL provides effective data enhancement by progressively improving the high-frequency components in the training data. During this process, PTL guides the training data distribution of these high-frequency components to approach that of the observations, while the low-frequency data is jointly updated under the physics-based constraints from FWI. Therefore, PTL implies a connection between the well-formulated physics-based FWI and the recently significantly developed deep learning approach. The low-frequency data extrapolated by PTL plays an important role in retrieving large-scale and low-resolution information about the subsurface structures. Especially when strong reflectors like salt domes and carbonates are present in the model, since most of the high-frequency energy cannot penetrate these reflectors, the low-frequency data could reveal the otherwise shielded sub-salt structures. Similar to other supervised learning approaches, PTL can easily embed prior knowledge into the subsurface structure by customizing how to generate training data. Moreover, compared to end-to-end supervised learning approaches, PTL shows superior performance in most cases, including the challenging reflection-dominant FWI and the complicated field data experiments, because PTL leverages the physics information solved from FWI to train the deep neural network. Therefore, PTL is suitable when the amount of training data is limited. Meanwhile, PTL is proven to be flexible by various experiments in different data configurations, such as different data acquisitions, sensor arrangements, and data pre-processing approaches. The method is validated by both the synthetic data and the field data, where the FWI results are obviously improved with the enhanced data.

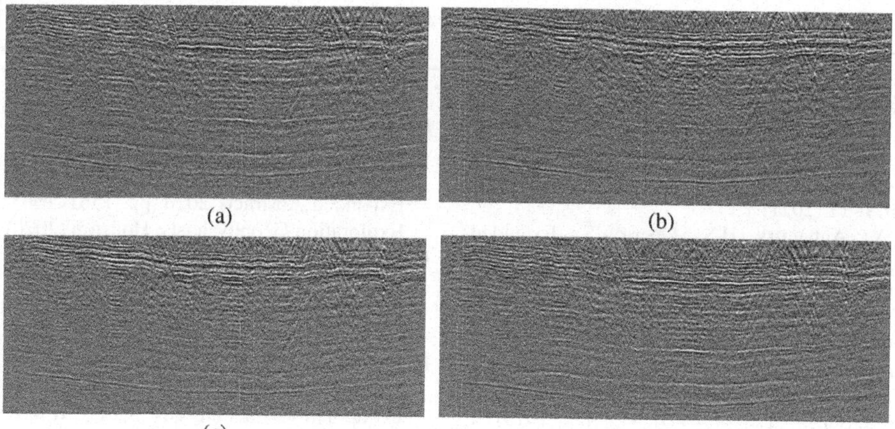

Fig. 2.29 Time domain imaging result of Fig. 2.27(**a**)–(**d**), respectively

References

Aghamiry, H.S., Gholami, A., Operto, S.: Dual variable compression: remedy the memory issue of full-space approaches for full-waveform inversion. In: First International Meeting for Applied Geoscience & Energy, pp. 772–776. Society of Exploration Geophysicists, Houston (2021)

Alkhalifah, T., Choi, Y.: From tomography to full-waveform inversion with a single objective function. Geophysics **79**(2), R55–R61 (2014)

Anagaw, A.Y., Sacchi, M.D.: Full waveform inversion with simultaneous sources using the full Newton method. In: 2012 SEG Annual Meeting. OnePetro (2012)

Asnaashari, A., Brossier, R., Garambois, S., Audebert, F., Thore, P., Virieux, J.: Time-lapse imaging using regularized FWI: a robustness study. In: SEG Technical Program Expanded Abstracts 2012, pp. 1–5. Society of Exploration Geophysicists, Houston (2012)

Bader, M., Clapp, R., Biondi, B.: Low frequency denoising using high frequency prediction error filters. In: SEG International Exposition and Annual Meeting, pp. 4500–4504. OnePetro, Richardson (2019)

Bednar, J.B., Shin, C., Pyun, S.: Comparison of waveform inversion, part 2: phase approach. Geophys. Prospect. **55**(4), 465–475 (2007)

Bénar, C.G., Papadopoulo, T., Clerc, M.: Topography-time-frequency atomic decomposition for event-related M/EEG signals. In: 2007 29th Annual International Conference of the IEEE Engineering in Medicine and Biology Society, pp. 5461–5464. IEEE, Piscataway (2007)

Brenders, A., Dellinger, J., Ahmed, I., Díaz, E., Gherasim, M., Jin, H., Vyas, M., Naranjo, J.: The wolfspar experience with low-frequency seismic source field data: motivation, processing, and implications. Lead. Edge **41**(1), 9–18 (2022)

Chen, R., Tran, K.T.: 2d Gauss-Newton full waveform inversion of SH- and Love-waves in the time domain. J. Appl. Geophys. **191**, 104363 (2021)

Chi, B., Dong, L., Liu, Y.: Full waveform inversion method using envelope objective function without low frequency data. J. Appl. Geophys. **109**, 36–46 (2014)

Chiu, S.K.: Low-frequency extrapolation using higher-frequency multi-channel prediction filters. In: SEG Technical Program Expanded Abstracts 2020, pp. 2835–2839. Society of Exploration Geophysicists, Houston (2020)

Chiu, S.K., Duggan, P., Jiang, Y.: Bandwidth extension using 3D structure preserving sparse-blind deconvolution. In: SEG Technical Program Expanded Abstracts 2018, pp. 4181–4185. Society of Exploration Geophysicists, Houston (2018)

Chopra, S., Castagna, J., Portniaguine, O.: Thin-bed reflectivity inversion. In: SEG Technical Program Expanded Abstracts 2006, pp. 2057–2061. Society of Exploration Geophysicists, Houston (2006)

Dondurur, D.: Acquisition and Processing of Marine Seismic Data. Elsevier, Amsterdam (2018)

Dong, S., Han, L., Hu, Y., Yin, Y.: Full waveform inversion based on a local traveltime correction and zero-mean cross-correlation-based misfit function. Acta Geophys. **68**(1), 29–50 (2020)

Fabien-Ouellet, G.: Low-frequency generation and denoising with recursive convolutional neural networks. In: SEG Technical Program Expanded Abstracts 2020, pp. 870–874. Society of Exploration Geophysicists, Houston (2020)

Fang, J., Zhou, H., Elita Li, Y., Zhang, Q., Wang, L., Sun, P., Zhang, J.: Data-driven low-frequency signal recovery using deep-learning predictions in full-waveform inversion. Geophysics **85**(6), A37–A43 (2020)

Fu, H., Ma, M., Han, B.: An accelerated proximal gradient algorithm for source-independent waveform inversion. J. Appl. Geophy. **177**, 104030 (2020)

Gao, Z., Hu, S., Li, C., Chen, H., Jiang, X., Pan, Z., Gao, J., Xu, Z.: A deep-learning-based generalized convolutional model for seismic data and its application in seismic deconvolution. IEEE Trans. Geosci. Remote Sens. **60**, 1–17 (2021)

Gholami, A., Aghamiry, H.S., Operto, S.: Extended-space full-waveform inversion in the time domain with the augmented Lagrangian method. Geophysics **87**(1), R63–R77 (2022)

Ha, J., Shin, S., Shin, C., Chung, W.: Time domain full waveform inversion of seismic data for the east sea, Korea, using a pseudo-hessian method with source estimation. J. Seismic Explor. **24**(5), 419–437 (2015)

He, Q., Han, B.: Accelerating full waveform inversion using HSS solver and limited memory conjugate gradient method. J. Appl. Geophys. **159**, 83–92 (2018)

Hu, W.: FWI without low frequency data-beat tone inversion. In: SEG Technical Program Expanded Abstracts 2014, pp. 1116–1120. Society of Exploration Geophysicists, Houston (2014)

Hu, W.: Beat tone FWI for cycle-skipping suppression–potentials and challenges. In: 2015 Workshop: Depth Model Building: Full-Waveform Inversion, Beijing, China, 18–19 June 2015, pp. 65–66. Society of Exploration Geophysicists, Houston (2015)

Hu, W., Chen, J., Liu, J., Abubakar, A.: Retrieving low wavenumber information in FWI: an overview of the cycle-skipping phenomenon and solutions. IEEE Signal Process. Mag. **35**(2), 132–141 (2018)

Hu, W., Jin, Y., Wu, X., Chen, J.: Physics-guided self-supervised learning for low frequency data prediction in FWI. In: SEG Technical Program Expanded Abstracts 2020, pp. 875–879. Society of Exploration Geophysicists, Houston (2020)

Hu, W., Jin, Y., Wu, X., Chen, J.: Progressive transfer learning for low-frequency data prediction in full-waveform inversion. Geophysics **86**(4), R369–R382 (2021)

Irabor, K., Warner, M.: Reflection FWI. In: 2016 SEG International Exposition and Annual Meeting, pp. 1136–1140. Society of Exploration Geophysicists, Houston (2016)

Jin, Y., Hu, W., Wu, X., Chen, J.: Learn low wavenumber information in FWI via deep inception based convolutional networks. In: 2018 SEG International Exposition and Annual Meeting. OnePetro, Richardson (2018)

Jin, Y., Zi, Y., Hu, W., Wu, X., Chen, J.: A deep learning enhanced full waveform inversion scheme. In: 2021 International Applied Computational Electromagnetics Society Symposium (ACES), pp. 1–4. IEEE, Piscataway (2021)

Jin, Y., Hu, W., Wang, S., Zi, Y., Wu, X., Chen, J.: Efficient progressive transfer learning for full-waveform inversion with extrapolated low-frequency reflection seismic data. IEEE Trans. Geosci. Remote Sens. **60**, 1–10 (2022a)

Jin, Y., Zi, Y., Hu, W., Hu, Y., Wu, X., Chen, J.: A robust learning method for low-frequency extrapolation in GPR full waveform inversion. IEEE Geosci. Remote Sens. Lett. **19**, 1–5 (2022b)

Kazei, V., Ovcharenko, O., Alkhalifah, T.: Velocity model building by deep learning: from general synthetics to field data application. In: SEG Technical Program Expanded Abstracts 2020, pp. 1561–1565. Society of Exploration Geophysicists, Houston (2020)

Kim, Y., Shin, C., Calandra, H., Min, D.J.: An algorithm for 3d acoustic time-Laplace-Fourier-domain hybrid full waveform inversion. Geophysics **78**(4), R151–R166 (2013)

Liang, C., Castagna, J., Torres, R.Z.: Tutorial: spectral bandwidth extension – invention versus harmonic extrapolation. Geophysics **82**(4), W1–W16 (2017). https://doi.org/10.1190/geo2015-0572.1

Lin, Z., Guo, R., Li, M., Abubakar, A., Zhao, T., Yang, F., Xu, S.: Low-frequency data prediction with iterative learning for highly nonlinear inverse scattering problems. IEEE Trans. Microw. Theory Tech. **69**(10), 4366–4376 (2021)

Liu, Y., Teng, J., Xu, T., Wang, Y., Liu, Q., Badal, J.: Robust time-domain full waveform inversion with normalized zero-lag cross-correlation objective function. Geophys. J. Int. **209**(1), 106–122 (2017)

Liu, X., Liu, S., Fu, L.: Full waveform inversion of cross-hole radar data using envelope objective function. In: 2018 17th International Conference on Ground Penetrating Radar (GPR), pp. 1–5. IEEE, Piscataway (2018)

Liu, Y., Liu, B., Huang, J., Wang, J., Chen, H., Liu, W.: Layered low-frequency extrapolation with deep learning in full-waveform inversion. In: Tenth International Symposium on Precision Mechanical Measurements, vol. 12059, pp. 367–374. SPIE, California (2021)

Mache, S., Pokala, P.K., Rajendran, K., Seelamantula, C.S.: Nuspan: a proximal average network for nonuniform sparse model–application to seismic reflectivity inversion. arXiv:2105.00003 (2021)

Maharramov, M., Baumstein, A.I., Tang, Y., Routh, P.S., Lee, S., Lazaratos, S.K.: Time-domain broadband phase-only full-waveform inversion with implicit shaping. In: SEG Technical Program Expanded Abstracts 2017, pp. 1297–1301. Society of Exploration Geophysicists, Houston (2017)

Métivier, L., Brossier, R., Virieux, J., Operto, S.: Full waveform inversion and the truncated Newton method. SIAM J. Sci. Comput. **35**(2), B401–B437 (2013)

Métivier, L., Brossier, R., Operto, S., Virieux, J.: Acoustic multi-parameter FWI for the recons54truction of P-wave velocity, density and attenuation: preconditioned truncated Newton approach. In: SEG Technical Program Expanded Abstracts 2015, pp. 1198–1203. Society of Exploration Geophysicists, Houston (2015)

Oh, J.W., Alkhalifah, T.: Full waveform inversion using envelope-based global correlation norm. Geophys. J. Int. **213**(2), 815–823 (2018)

Ovcharenko, O., Kazei, V., Peter, D., Zhang, X., Alkhalifah, T.: Low-frequency data extrapolation using a feed-forward ANN. In: 80th EAGE Conference and Exhibition 2018, pp. 1–5. European Association of Geoscientists & Engineers, Utrecht (2018)

Ovcharenko, O., Kazei, V., Kalita, M., Peter, D., Alkhalifah, T.: Deep learning for low-frequency extrapolation from multioffset seismic data. Geophysics **84**(6), R989–R1001 (2019)

Ovcharenko, O., Kazei, V., Peter, D., Alkhalifah, T.: Transfer learning for low frequency extrapolation from shot gathers for FWI applications. In: 81st EAGE Conference and Exhibition 2019, pp. 1–5. European Association of Geoscientists & Engineers, Utrecht (2019)

Ovcharenko, O., Kazei, V., Plotnitskiy, P., Peter, D., Silvestrov, I., Bakulin, A., Alkhalifah, T.: Extrapolating low-frequency prestack land data with deep learning. In: SEG Technical Program Expanded Abstracts 2020, pp. 1546–1550. Society of Exploration Geophysicists, Houston (2020)

Ovcharenko, O., Kazei, V., Peter, D., Alkhalifah, T.: Transferring elastic low frequency extrapolation from synthetic to field data. In: 82nd EAGE Annual Conference & Exhibition, pp. 1–5. European Association of Geoscientists & Engineers, Utrecht (2021)

Ovcharenko, O., Kazei, V., Peter, D., Silvestrov, I., Bakulin, A., Alkhalifah, T.: Dual-band generative learning for low-frequency extrapolation in seismic land data. In: First International Meeting for Applied Geoscience & Energy, pp. 1345–1349. Society of Exploration Geophysicists, Houston (2021)

Pan, W., Innanen, K.A., Margrave, G.F., Cao, D.: Efficient pseudo-gauss-newton full-waveform inversion in the τ-p domain. Geophysics **80**(5), R225–R14 (2015)

Pan, W., Innanen, K.A., Margrave, G.F., Fehler, M.C., Fang, X., Li, J.: Estimation of elastic constants for HTI media using Gauss-Newton and full-Newton multiparameter full-waveform inversion. Geophysics **81**(5), R275–R291 (2016)

Pan, W., Innanen, K.A., Liao, W.: Accelerating Hessian-free Gauss-Newton full-waveform inversion via l-BFGS preconditioned conjugate-gradient algorithm. Geophysics **82**(2), R49–R64 (2017)

Pérez, D.O., Velis, D.R., Sacchi, M.D.: High-resolution prestack seismic inversion using a hybrid FISTA least-squares strategy. Geophysics **78**(5), R185–R195 (2013)

Qiu, L., Chemingui, N., Zou, Z., Valenciano, A.: Full-waveform inversion with steerable variation regularization. In: 2016 SEG International Exposition and Annual Meeting. OnePetro, Richardson (2016)

Ren, Q.: Seismic acoustic full waveform inversion based on the steepest descent method and simple linear regression analysis. J. Appl. Geophys. **203**, 104686 (2022)

Schäfer, M., Groos, L., Forbriger, T., Bohlen, T.: On the effects of geometrical spreading corrections for a 2D full waveform inversion of recorded shallow seismic surface waves. In: 74th EAGE Conference and Exhibition Incorporating EUROPEC 2012, pp. cp–293. European Association of Geoscientists & Engineers, Utrecht (2012)

Sheng, J., Mao, J., Liu, F., Hart, M.: A robust phase-only reflection full waveform inversion with multi-channel local correlation. In: EAGE 2020 Annual Conference & Exhibition Online, pp. 1–5. European Association of Geoscientists & Engineers, Utrecht (2020)

Shin, C., Ho Cha, Y.: Waveform inversion in the Laplace–Fourier domain. Geophys. J. Int. **177**(3), 1067–1079 (2009)

Smith, M., Perry, G., Stein, J., Bertrand, A., Yu, G.: Extending seismic bandwidth using the continuous wavelet transform. First Break **26**(6) (2008)

Sun, H., Demanet, L.: Low frequency extrapolation with deep learning. In: SEG Technical Program Expanded Abstracts 2018, pp. 2011–2015. Society of Exploration Geophysicists, Houston (2018)

Sun, H., Demanet, L.: Extrapolated full waveform inversion with convolutional neural networks. In: SEG Technical Program Expanded Abstracts 2019, pp. 4962–4966. Society of Exploration Geophysicists, Houston (2019)

Sun, H., Demanet, L.: Elastic full-waveform inversion with extrapolated low-frequency data. In: SEG Technical Program Expanded Abstracts 2020, pp. 855–859. Society of Exploration Geophysicists, Houston (2020a)

Sun, H., Demanet, L.: Extrapolated full-waveform inversion with deep learning. Geophysics **85**(3), R275–R288 (2020b)

Sun, H., Demanet, L.: Deep learning for low frequency extrapolation of multicomponent data in elastic FWI. IEEE Trans. Geosci. Remote Sens. (2021)

Sun, Y., Araya-Polo, M., Williamson, P.: Data characterization and transfer learning for DL-driven velocity model building. In: First International Meeting for Applied Geoscience & Energy, pp. 1475–1479. Society of Exploration Geophysicists, Houston (2021)

Virieux, J., Operto, S.: An overview of full-waveform inversion in exploration geophysics. Geophysics **74**(6), WCC1–WCC26 (2009)

Virieux, J., Asnaashari, A., Brossier, R., Métivier, L., Ribodetti, A., Zhou, W.: An introduction to full waveform inversion. In: Encyclopedia of Exploration Geophysics, pp. R1–1. Society of Exploration Geophysicists, Houston (2017)

Wang, M., Xu, S., Zhou, H.: Self-supervised learning for low frequency extension of seismic data. In: SEG Technical Program Expanded Abstracts 2020, pp. 1501–1505. Society of Exploration Geophysicists, Houston (2020)

Wang, P., Zhang, Z., Mei, J., Lin, F., Huang, R.: Full-waveform inversion for salt: a coming of age. Lead. Edge **38**(3), 204–213 (2019)

Xie, X.B.: Recover certain low-frequency information for full waveform inversion. In: 2013 SEG Annual Meeting. OnePetro, Utrecht (2013)

Xue, Z., Zhu, H., Fomel, S.: Full-waveform inversion using seislet regularization. Geophysics **82**(5), A43–A49 (2017)

Yan, Y.: Two-dimensional love-wave full waveform inversion in time domain—combine DFP algorithm and steepest descent algorithm. In: International Conference on Engineering Geophysics, Al Ain, United Arab Emirates, 9–12 October 2017, pp. 120–123. Society of Exploration Geophysicists, Houston (2017)

Yilmaz, Ö.: Seismic data analysis: processing, inversion, and interpretation of seismic data. Society of Exploration Geophysicists, Houston (2001)

Zhang, R., Castagna, J.: Seismic sparse-layer reflectivity inversion using basis pursuit decomposition. Geophysics **76**(6), R147–R158 (2011)

Zhang, P., Han, L., Xu, Z., Zhang, F., Wei, Y.: Sparse blind deconvolution based low-frequency seismic data reconstruction for multiscale full waveform inversion. J. Appl. Geophys. **139**, 91–108 (2017)

Deep Learning for Seismic Deblending

3

3.1 Introduction

3.1.1 Simultaneous Source Acquisition

The significance of seismic reflection surveys in geophysical exploration cannot be understated, as they generate acoustic portrayals of the earth's interior. Traditional seismic data acquisition emphasizes the individual collection of responses from separate seismic sources, referred to as 'shots'. To avoid interference among these seismic responses, substantial time intervals[1] ranging from a few seconds to half a minute are required between successive shots. However, this method of using separate sources leads to increased costs due to the lengthy survey process and associated efforts. Consequently, the pursuit of greater efficiency inspired the emergence of simultaneous-source acquisition, otherwise known as blended acquisition (Beasley et al., 1998; Berkhout, 2008). A denser blended acquisition is attained in a shorter survey period by triggering multiple shots within a relatively short time interval using a random time-dithering technique, as illustrated in Fig. 3.1.

In the context of conventional (separate-source) acquisition, a seismic trace is collected as a time series that captures the response to seismic sources at a specific receiver point. Seismic data in the time domain can be represented as a multidimensional array $P_{ijk} = P(x_i^{\mathrm{s}}, x_j^{\mathrm{r}}, t_k)$, where x_i^{s} corresponds to the ith source position, x_j^{r} signifies the jth receiver location, and t_k is the kth time sample. The array P exhibits a shape of $n_{\mathrm{s}} \times n_{\mathrm{r}} \times n_{\mathrm{t}}$, where n_{s} designates the number of separate sources, n_{r} denotes the receiver count, and n_{t} represents the time series length. In the case of blended acquisition, we follow the formulation of (Wapenaar et al., 2012). A group of sources is denoted as σ_l, where lth represents the subset index of this group among all sources. And the source location is defined as x_i^{s}. The sources in subset σ_l are triggered within a relatively short and randomly distributed time intervals[2] τ_{li} (dithering). The blended data can be defined as

$$P_{ljk}^{\mathrm{b}} = P^{\mathrm{b}}(\sigma_l, x_j^{\mathrm{r}}, t_k) = \begin{cases} \sum_{x_i^{\mathrm{s}} \in \sigma_l} P(x_i^{\mathrm{s}}, x_j^{\mathrm{r}}, t_k - \tau_{li}) & \text{for } t_k > \tau_{li}, \\ 0 & \text{else.} \end{cases} \tag{3.1}$$

Given the number of source groups n_{b}, the dimension of P^{b} is $(n_{\mathrm{b}} \times n_{\mathrm{r}} \times n_{\mathrm{t}})$. The temporal Fourier transforms of P_{ijk} and P_{ljk}^{b} are represented as $\Pi_{ijk} = \Pi(x_i^{\mathrm{s}}, x_j^{\mathrm{r}}, \omega_k)$ and

[1] A time span that exceeds the duration needed to capture reflections from the deepest reflectors of interest (Wapenaar et al., 2012).

[2] The quantization of these delay times matches that of the time steps t_k.

S. Wang et al., *Deep Learning for Seismic Data Enhancement and Representation*, Advances in Oil and Gas Exploration & Production, https://doi.org/10.1007/978-3-031-75745-7_3

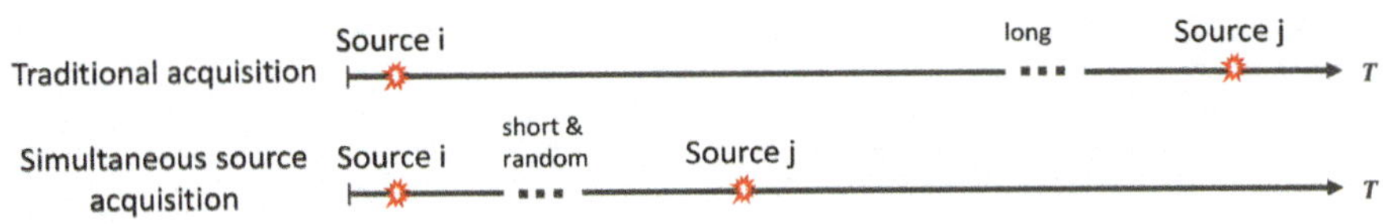

Fig. 3.1 Illustration of simultaneous source acquisition. Compared to conventional separate source acquisition, the intervals between sources are much shorter

$\Pi^{\mathrm{b}}_{ljk} = \Pi^{\mathrm{b}}(\sigma_l, x^{\mathrm{r}}_j, \omega_k)$, respectively. In this case ω_k denotes the kth angular frequency and Π, Π^{b} posses the same dimensions as P, P^{b}, respectively. Therefore, the blending procedure shown in Eq. (3.1) can be formulated in the frequency domain as phase-shifted terms

$$\Pi^{\mathrm{b}}(\sigma_l, x^{\mathrm{r}}_j, \omega_k) = \sum_{x^{\mathrm{s}}_i \in \sigma_l} \mathrm{e}^{-\mathrm{i}\omega_k \tau_{li}} \, \Pi(x^{\mathrm{s}}_i, x^{\mathrm{r}}_j, \omega_k), \tag{3.2-1}$$

$$\Gamma_{lik} = \begin{cases} \mathrm{e}^{-\mathrm{i}\omega_k \tau_{li}} & \text{for } x^{\mathrm{s}}_i \in \sigma_l, \\ 0 & \text{else.} \end{cases}, \tag{3.2-2}$$

$$\Pi^{\mathrm{b}}(\sigma_l, x^{\mathrm{r}}_j, \omega_k) = \sum_{i} \Gamma_{lik} \, \Pi(x^{\mathrm{s}}_i, x^{\mathrm{r}}_j, \omega_k). \tag{3.2-3}$$

For a simplified representation, all indices are omitted except for the time and frequency index in an array to denote the entire range of index selection. As an illustration, Π_k represents the matrix extracted from the three-dimensional data Π with particular ω_k. Consequently, the blending procedure outlined in Eq. (3.2-3) is reformulated through matrix multiplication

$$\Pi^{\mathrm{b}}_k = \Gamma_k \, \Pi_k. \tag{3.3}$$

Simultaneous shooting greatly enhances survey productivity, yielding high-density data at a lower expense. Despite these financial advantages, strong interferences between different shots pose a challenge for traditional seismic data processing workflows. As a result, deblending is needed to eliminate these interferences, dividing simultaneous-source acquisition data into individual-source responses to support post-acquisition evaluation. Therefore, the deblending problem is defined as determining Π_k given Π^{b}_k and Γ_k.

Deblending presents an under-determined challenge, and a pseudo-deblending solution (Berkhout, 2008; Mahdad et al., 2011) is often firstly achieved through the least-squares approach. Letting Γ^{H}_k represent the conjugate transpose of Γ_k, the pseudo-deblended solution Π^{LS}_k is calculated as

$$\Pi^{\mathrm{LS}}_k = \Gamma^{\mathrm{H}}_k (\Gamma_k \Gamma^{\mathrm{H}}_k)^{-1} \Pi^{\mathrm{b}}_k. \tag{3.4}$$

With a common assumption that every source is only triggered once and will not be extended across multiple source groups, i.e. $\sigma_m \cap \sigma_n = \emptyset$ for all indices $m \neq n$, Γ_k features mutually orthogonal rows, the least-squares solution can be further simplified as $\Pi^{\mathrm{LS}}_k = \widehat{\Gamma}^{\mathrm{H}}_k \Pi^{\mathrm{b}}_k$, where each row of $\widehat{\Gamma}_k$ is obtained by the division of the corresponding row in Γ_k and its row-wise squared norm. Usually, the pseudo-deblending process refers to the multiplication $\widehat{\Gamma}^{\mathrm{H}}_k \Pi^{\mathrm{b}}_k$ regardless of whether the orthogonal-rows condition is satisfied for Γ_k. The pseudo-deblended data is then denoted as

$$\widehat{\Pi}_k = \widehat{\Gamma}^{\mathrm{H}}_k \Pi^{\mathrm{b}}_k. \tag{3.5}$$

Unfortunately, the pseudo-deblending procedure does not effectively eliminate undesired interferences, i.e. the blending noises. Instead, it primarily reshapes the data array Π^{b}_k back to its original dimensions of $(n_{\mathrm{s}} \times n_{\mathrm{r}})$. This process involves replicating and relocating the blended shot gathers to their ordered positions, in accordance with the implemented blending strategy, followed by reversing the dithering delays. After applying pseudo-deblending and performing an inverse Fourier transform, the desired source signals within the common-receiver, common-midpoint, and common-offset domains are prop-

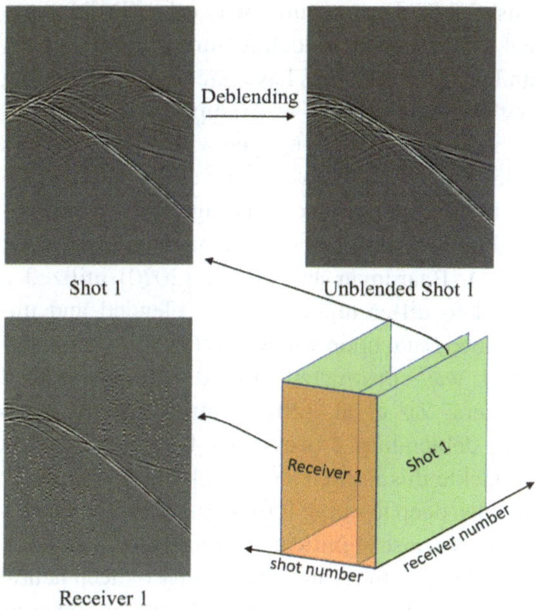

Fig. 3.2 The bottom-right section of the figure presents the 3D volume of the pseudo-deblended seismic data. In the top row, two images compare the pseudo-deblended data with the corresponding unblended data in the common-shot domain. The interference, i.e. blending noise from another sources is coherent. In contrast, the blending noise in the common receiver domain (bottom-left) manifests as random noises

erly aligned and continuous. However, interferences from other sources manifest as incoherent noise, as demonstrated in Fig. 3.2.

At this juncture, the deblending process transitions into a specialized denoising challenge where the unblended signals and blending noises exhibit identical amplitude distributions but differ significantly in coherency. The definition $\widetilde{\Pi}_k$ for this denoising problem can be expressed as

$$\widetilde{\Pi}_k = \arg\min_z \frac{1}{2}\|\widehat{\Pi}_k - \widehat{\Gamma}_k^{\mathrm{H}}\Gamma_k z\|_2^2 + \mathcal{R}(z), \quad (3.6)$$

where $\mathcal{R}$ denotes a regularization operator that constrains the optimization to a noise-suppressed result, while z represents the desired deblending solution. The characteristics of blending noise significantly depend on the selected blending schemes. Factors such as the randomness of dithering time, the number of blended shots, spatial distance between blended shots, source wavelet spectrum, and others during acquisition configurations can dramatically impact the extent of blending noise, thus testing the capabilities of deblending algorithms.

3.1.2 Conventional Approaches for Seismic Deblending

In recent studies of deblending techniques, the emphasis has been on either diminishing blending noise through filter-based approaches (Huo et al., 2012) or directly separating sources by solving the inversion problem (Moore et al., 2008). The majority of blending-noise reduction methods depend on the incoherence of blending noises in common-receiver, common-midpoint, and common-offset domains after pseudo-deblending by applying random dithering to the source-group during blended acquisition (Mahdad et al., 2011). Huo et al. (2012) proposed to filter the incoherent noise in the common receiver domain by a directional-sensitive median filter. Chen (2015) adopted a similar concept by allowing the median filter to vary along the seismic events, which further expanded the discussions for filtering-based techniques. An iterative coherency improvement framework was introduced by Mahdad et al. (2011, 2012) which adopted a filter followed by a thresholding step in the $f-k$ domain to estimate the noise, enabling a preciser signal reconstruction. The incorporation of median filter and $f-x$ projection filter was proposed by Peng et al. (2013) to estimate coherent signals from contaminated signals in an iterative approach. Apart from the filtering method, sparsity of seismic signals was also considered an important feature and many works have proved the effectiveness of adopting the sparsity constraints for random noise removal. Chen et al. (2014) improved the sparsity of seismic gathers in the seislet domain with thresholding. Zhang and Wang (2015) leveraged the data sparsity assumption in the Tau-p domain and incorporated it into the inversion process. Other strategies involve investigating the characteristic of low-rankness for seismic data. Specifically, the singular spectra analysis (SSA) was adopted to estimate and reduce the rank for seismic data by Cheng and Sacchi (2013) and improved the coherency of signals in blended seismic gathers. Zu et al. (2017) incorporated the thresholding method into SSA and iteratively suppress the rank of seismic gathers, as well as the random noises within. In another way, Wapenaar et al. (2012) proposed to utilize a band-limited point-spread function throughout the inversion procedure. Generally, iterative noise elimination approaches (Abma et al., 2015; Bahia et al., 2020) are favored over non-iterative ones. The coherency in seismic gathers is progressively improved by sophisticated filters combined with regularization terms in specific transformed domains.

While these studies have achieved a certain degree of success, several challenges persist to leave space for improvement. Firstly, as seismic waves propagate, energy diffusion causes early-arrival signals to be significantly stronger than late arrivals when there is a considerable time delay between shots. This results in undesired and potent blending noises that can completely mask the signal in some regions, rendering it irrecoverable by time-space filters which often demand a high signal-to-noise ratio (SNR). Although iterative deblending algorithms can partially address this issue, their performances remain inadequate for data with intricate patterns. Secondly, tuning hyper-parameters for these methods, such as the size of filters, threshold in sparsity constraint, rank estimation and selection, etc. is not a simple task. It is crucial to find a parameter for properly balancing the noise suppression and valuable signals preservation. Furthermore, these parameters are not universally applicable across different samples, necessitating time-consuming and repetitive hyper-parameter tuning by skilled researchers during exhaustive iterations.

3.1.3 Challenge of Supervised Learning Scheme for Seismic Deblending

In recent times, deep learning (LeCun et al., 2015) has garnered considerable interest across various fields. Deep neural networks (DNNs) excel at identifying hidden features within signals, and numerous studies have integrated deblending with deep learning. For example, a U-net (Ronneberger et al., 2015) merged with residual blocks (He et al., 2016) was employed by Richardson and Feller to estimate blending noises from several neighboring gathers (Richardson and Feller, 1907). Baardman and Hegge (2020) utilized a CNN to differentiate between blended and unblended data, upon which another regression network was constructed to predict the deblended gathers. Zu et al. (2020) introduced an iterative deblending process that leverages a CNN to tackle this issue. Despite achieving promising results, deep learning for seismic deblending continues to pose significant challenges.

The robustness and adaptability of deep neural networks (DNNs) continue to be major concerns in practical applications, especially when data patterns vary across different fields. Moreover, obtaining fully labeled training datasets for seismic data processing poses a significant challenge, particularly for deblending issues. Therefore, it is essential to develop a reliable deblending model that can generalize effectively, perform robustly, and provide consitently good results across diverse data volumes collected under different conditions.

3.2 Physics Augmented Deep Learning for Seismic Deblending

3.2.1 Iterative Training Adaptation for Reliable Deblending

For the seismic deblending task, supervised learning methods are heavily reliant on an extensive collection of labeled data. However, when the underlying distributions of a new dataset and the training dataset differ significantly, the performance of the trained network may be suboptimal. To mitigate the challenge posed by the scarcity of dependable datasets, we introduce an innovative iterative training approach that utilizes training datasets derived from a progressively evolving

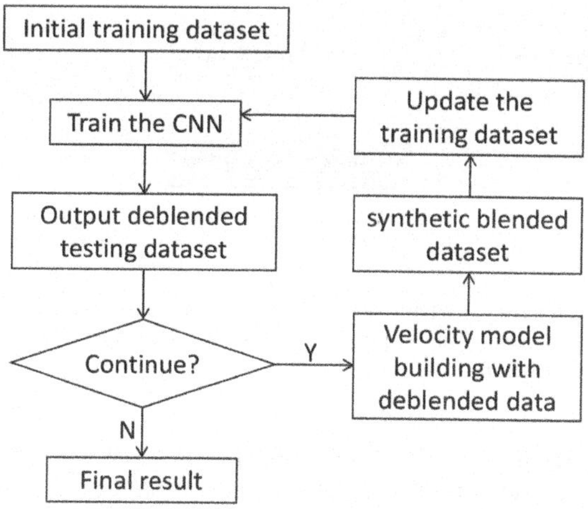

Fig. 3.3 Diagram of the physics-augmented training workflow. The process begins with training on an initial dataset, after which a convolutional neural network (CNN) generates a preliminary deblended prediction for the testing dataset. This prediction undergoes inversion and simulation procedures, producing a new training set. The CNN is then iteratively updated with this new training data, repeating the cycle until convergence is achieved. This iterative approach ensures that the network aligns with the underlying physics, thereby improving its performance on the testing dataset

velocity model. This model ultimately converges to a highly representative one as the learning iterations advance. The schematic representation of this cutting-edge deblending workflow, which employs an iteratively trained deep neural network (DNN) with physics-augmented datasets, can be observed in Fig. 3.3.

The physics-augmented workflow is designed to identify and eliminate incoherency stemming from simultaneous-source interference in the deblending process. The workflow begins with a synthesized blended data which is produced using a random velocity model that matches the acquisition parameters of the actual testing data. Since the training data is synthetic, both the unblended seismic (ground truth) data and its blended counterpart (input) are known and form the initial training dataset. After trained with the synthetic data in the first round, the network is subsequently used for deblending the target testing dataset. At this step, although not optimal, the deblending results are utilized to refine the velocity model, which is subsequently used to replace the initial random velocity model and generate a new synthetic training dataset for further network training. Through iterative refinement, the network's deblending accuracy can improve as the velocity model approaches the target model and the synthesized training data becomes more representative of the real testing dataset.

3.2.2 Network Structure and Loss Design

Various convolutional neural networks (CNNs) have been constructed for a multitude of tasks. Among them, the encoder-decoder CNN, such as U-net (Ronneberger et al., 2015), has been widely implemented across numerous applications. In our work, we follow the construction of the U-Net structure, replacing the last activation layer by a leaky-ReLU activation. It is employed to directly predict the clean deblended data from the noisy pseudo-deblended inputs. There are four stages of down-sampling blocks in the encoding section to process the input and extract seismic attributes from different levels of resolution. Each encoder level consists of two consecutive convolutional layers with a kernel size of (3 × 3) and each of them is followed by a ReLU activation layer. A max-pooling layer is used after each encoder block. Two additional convolutional layers are employed in the middle section. There are four corresponding up-sampling blocks to restore the seismic details and transform the feature map to its original dimension. Skip connections are adopted to directly concatenate the encoding block outputs and decoding block inputs correspondingly.

Figure 3.4 illustrates the training workflow for a single iteration. The network input is the pseudo-deblended data, generated by copying and reverting the time delays for each shot (Mahdad et al., 2011). The data volume is sorted in the common receiver domain where the unwanted interference becomes incoherent, referred to as blending noises, which the network aims to eliminate. The network output consists

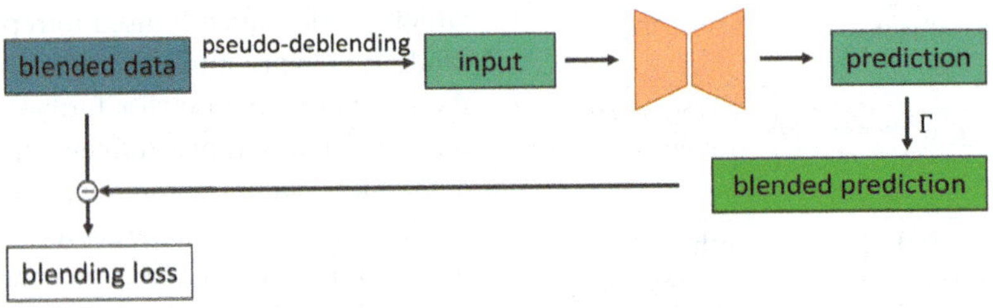

Fig. 3.4 Definition of the blending loss. The network gives a deblended prediction, with which a blended prediction is generated using the blending matrix Γ in the frequency domain. The blending loss is computed by comparing the original blended data with the blended prediction

of predicted signals for separate sources. The objective function for the training process can be formulated as

$$\mathcal{L} = w_{\text{MSE}}\mathcal{L}_{\text{MSE}} + w_{\text{blending}}\mathcal{L}_{\text{blending}}, \tag{3.7}$$

where the first term is defined as

$$\mathcal{L}_{\text{MSE}} = \frac{1}{m}\sum_{i=1}^{m} \| f_\theta(x_i) - y_i^* \|_2^2, \tag{3.8}$$

in which f_θ denotes the network model parameterized by the weight θ. x_i and y_i^* represent the pseudo-deblended input and unblended ground truth, respectively. An additional blending loss function

$$\mathcal{L}_{\text{blending}} = \frac{1}{m}\sum_{i=1}^{m} \| z_i - z_i^* \|_2^2, \tag{3.9}$$

where

$$z_i = \mathcal{F}^{-1}(\Gamma \cdot \mathcal{F}(f_\theta(x_i))), \tag{3.10}$$

is incorporated as a regularization term to ensure the energy is matched and the amplitudes are preserved, improving the generalization capability of the neural network. The blending loss measures the discrepancy between the predicted blending, obtained by applying the blending operator Γ to the predicted deblended data, and the actual blended observation. Here, $\mathcal{F}$ and $\mathcal{F}^{-1}$ symbolize the Fourier transform and inverse Fourier transform, respectively. Moreover, Γ represents the blending operator and z_i^* denotes the blended ground truth. This holistic approach ensures the integrity of the deblending process while maintaining the desired network performance.

3.2.3 Experimental Results

We synthesized two datasets with distinct velocity models, namely the Sigsbee and BP2004 for training and testing, respectively. There are 200 sources deployed for the training dataset sharing the same 200 receivers. For the testing dataset, 224 sources are configured with 224 receivers per source. The total record lasts 5 s with a sampling rate of 2 ms while the receiver spacing is 10 m for both datasets. A random sample of the training data in the common receiver domain is displayed in Fig. 3.5. The dataset is augmented to double the size by random left-and-right flipping in the training stage. As demonstrated in Fig. 3.6, blending noise is not effectively removed following the initial training round, owing to the significant differences in geological structure between the testing and training data. Moreover, the predicted amplitudes of coherent signals are inaccurate.

However, after employing the proposed physics-augmented iterative training algorithm, predictions exhibit substantial performance improvements after two additional learning iterations. As the process progresses, the velocity model more accurately reflects the true model due to the training dataset becoming more representative of the actual data. The result from the 3rd round of the iterative training workflow is illustrated in Fig. 3.7, whose comparison with the first-round result is presented in Fig. 3.8. It is distinct that the majority of the artifacts are

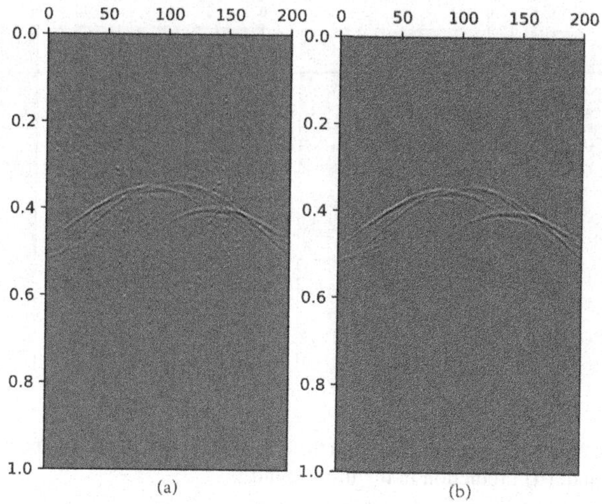

Fig. 3.5 Depiction of the initial training dataset, showcasing data features that are significantly different from those of the testing data in Fig. 3.2. (**a**) Input pseudo deblended data; (**b**) Ground truth

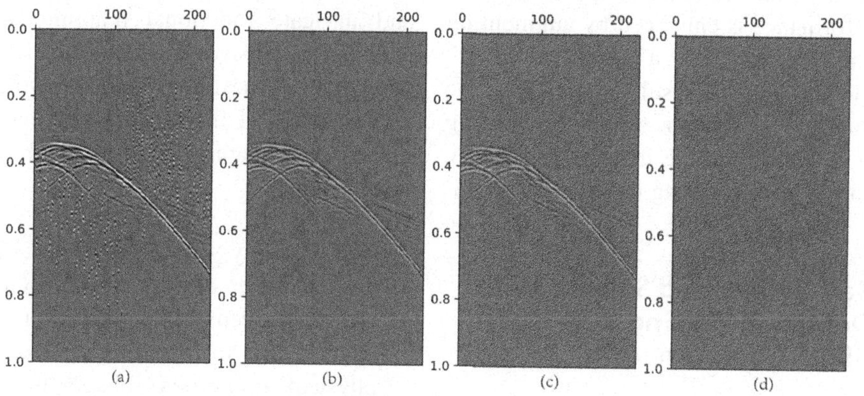

Fig. 3.6 Result of the first iteration training. (**a**), (**b**) Pseudo-deblended input and unblended ground truth of the testing data, respectively; (**c**) Prediction in the first round; (**d**) Residual between (**b**) and (**c**)

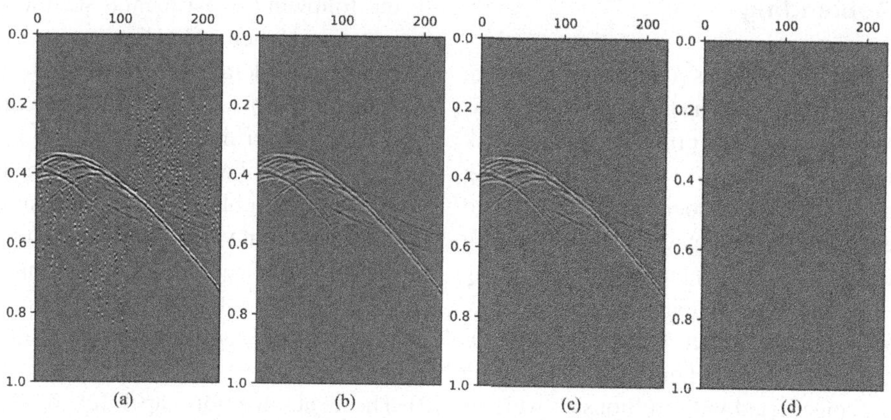

Fig. 3.7 Result of the testing data in the third round. (**a**), (**b**) Input and its ground truth, respectively; (**c**) Prediction from the network; (**d**) Residual between (**b**) and (**c**)

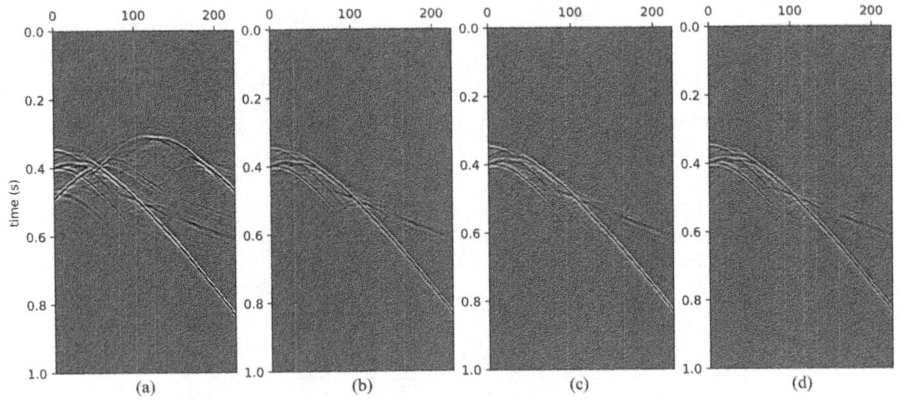

Fig. 3.8 Comparison of the results between the first and third iterations. (**a**), (**b**) Input and its ground truth, respectively; (**c**) Prediction in the first round; (**d**) Prediction in the third round

effectively removed, and the dominant events are predicted with accurate amplitudes. The DNN's performance is enhanced by augmenting the training dataset with a more "realistic" velocity model from the result in round 2. The distributional gap between the training and testing datasets is reduced with the iterative algorithm, thus mitigating the overfitting issue.

3.3 Self-Supervised Seismic Data Deblending Using a Blind-Trace Network

3.3.1 Self-Supervised Leaning Scheme for Seismic Deblending

Together with the physics-augmented iterative training algorithm, supervised learning has demonstrated significant promise in the task of seismic deblending, while addressing adaptability concerns associated with trained neural networks to some extent. These learning approaches train a deep neural network using prepared blended-unblended pairs before deploying it on the target dataset. Acting as sophisticated denoising filters, the deblending models are constructed with millions of weights specifically tailored to accommodate patterns within the training data. In comparison to conventional methods, deep neural networks excel at capturing both global and local features of seismic data, yielding a powerful and intricate functional mapping without the need for extensive hyperparameter fine-tuning. Nevertheless, these end-to-end networks continue to face some of the most prominent challenges that have plagued conventional deblending techniques.

(1) The efficacy of deblending performance declines when the signal-to-noise ratio (SNR) drops to extremely low levels for late arrivals. Specifically, the model does not utilize any physical properties related to deblending, causing the CNN to generate predictions based solely on visual patterns. As demonstrated in the following experimental section, there are instances where blending noise amplitudes can be considerably greater than the signals in specific regions. Under these circumstances, a purely discriminative model is unlikely to reconstruct coherent patterns from substantial noise. This highlights the limitations of relying solely on input visual patterns and underscores the importance of incorporating physical information into the deblending process to enhance performance in challenging scenarios.
(2) The generalization capability of the learned parameters can be uncertain when there are variations in the datasets, especially when the training and testing data share little resem-

blance. While the physics-augmented iterative training method addresses this issue to a certain degree, it still necessitates an initial synthetic dataset that ensures accurate convergence of the Full Waveform Inversion (FWI) model, which is essential in various situations. This emphasizes the necessity for not just enhancing the adaptability of the models, but investigating novel strategies that can adequately tackle the adaptiveness concern, perhaps by considering alternatives to the supervised learning scheme.

To address this issue, we propose an innovative self-supervised deep learning method that employs a blind-trace neural network. This approach allows for direct training on the target blended dataset without the need for any pre-labeled training data. Let $\mathcal{F}$ and $\mathcal{F}^{-1}$ denote the 1-D temporal Fourier transformation and its inverse, respectively. A denoising CNN $f_{\boldsymbol{\theta}}(\cdot)$ is parameterized by $\boldsymbol{\theta}$ and C symbolizes the concatenation operator that stacks arrays into a single elongated array along a certain axis, e.g. $\Pi = C\Pi_k$. The self-supervised blending loss function can be formulated in the absence of unblended data as

$$\mathcal{L}_\theta(\mathbf{x}, f_\theta(\mathbf{x})) = \frac{1}{N}\sum_{k=1}^{N}\|x^{(k)} - \mathcal{F}^{-1}C\Gamma_m^{\mathrm{H}}\Gamma_m \times (\mathcal{F}f_\theta(x^{(k)}))_m\|_1,$$
$$\mathbf{x} = \{x^{(1)}, \ldots, x^{(N)}\}, \tag{3.11}$$

where the input $\mathbf{x} = \mathcal{F}^{-1}C\Gamma_k^{\mathrm{H}}\Pi_k^{\mathrm{b}}$[3] denotes the pseudo-deblended common receiver gather. The deblended result is given by $f_{\boldsymbol{\theta}}(\mathbf{x})$, which is blended and pseudo-deblended again by $\Gamma_m^{\mathrm{H}}\Gamma_m$ using the known blending matrix Γ. The right side of the formula is called the blended prediction, which then subtracts the true pseudo-deblended input $\mathbf{x}$ to compute the mean absolute error. Utilizing this blending loss enables a pure self-supervised fashion for training the deblending model, assuming the availability of only the blended data and the blending matrix.

Nonetheless, due to the ill-posed nature of the problem, a conventional convolutional neural network composed of only vanilla convolutional layers (such as U-Net) does not ensure a valid deblending result when trained to minimize this blending loss. Instead, the model can easily learn to simply generate the pseudo-deblended data, which is almost identical to the input with a simple scaling operation, equivalent to the least square result. An example is illustrated in Fig. 3.9. In this example, a naive scaling of $1/2$ is naively applied to the input image including both the noise and signal,[4] which simply minimizes the blending loss to 0.

To circumvent this simplistic solution, we propose explicitly regularizing the receptive field of the network by incorporating a blind-trace constraint. This strategy is founded on two essential facts (1). Blending noise is trace-wise independent due to the incoherence property resulting from random dithering codes; (2). The coherent signal, in contrast, shows trace-wise dependence. Consequently, the coherent signal of a single trace can be theoretically reconstructed from its neighboring traces, while excluding itself in the input. As a result, the blending noise in a specific trace is strictly prevented from leaking from the input into the network prediction. Additionally, the model will naturally ignore any noise in the neighboring area due to its discontinuity, thereby predicting accurate signal values as the blending loss converges. This concept is depicted in Fig. 3.10.

3.3.2 Blind-Trace Network

A self-supervised denoising method is introduced by Krull et al. (2019) using a blind-spot training strategy, and Laine et al. (2019) further extend this concept with a blind-spot network which enables an efficient implementation for the blind-spot constraint. Inspired by the blind-spot network, we developed a blind-trace

[3] Here Γ_k^{H} is used instead of $\widehat{\Gamma}_k^{\mathrm{H}}$ to ensure that the amplitudes of both signal and noise in the images are preserved in the original scale, and not be divided by the blending factor.

[4] The blending factor is 2 in this case, i.e. every two shots are blended.

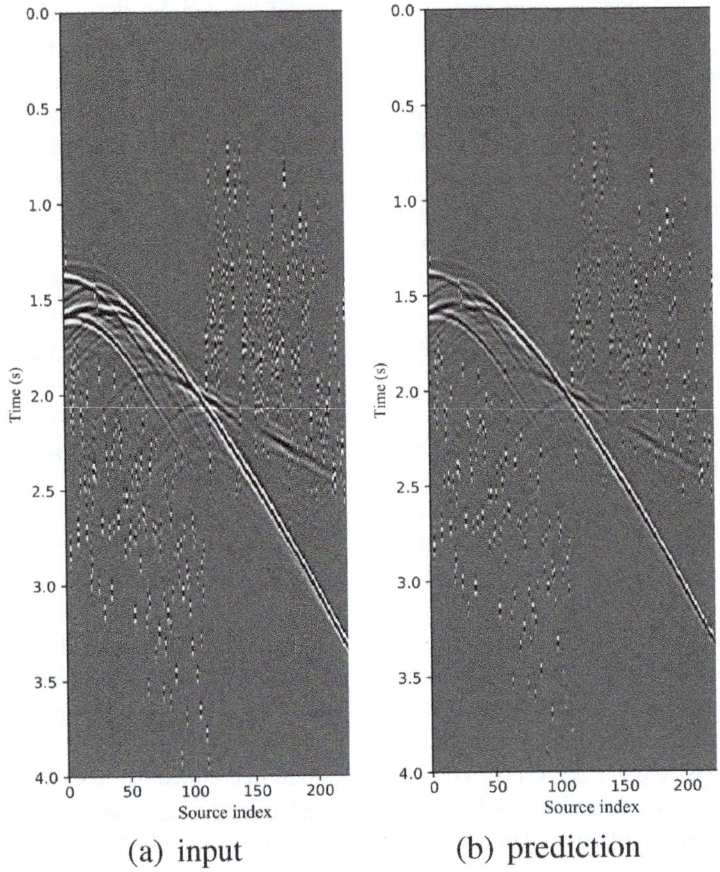

Fig. 3.9 Issue with directly training a conventional CNN (U-Net) using the blending loss Eq. (3.11). (**a**) Noisy input; (**b**) Prediction by a fully converged U-net using the same configuration in Sect. 3.2.2

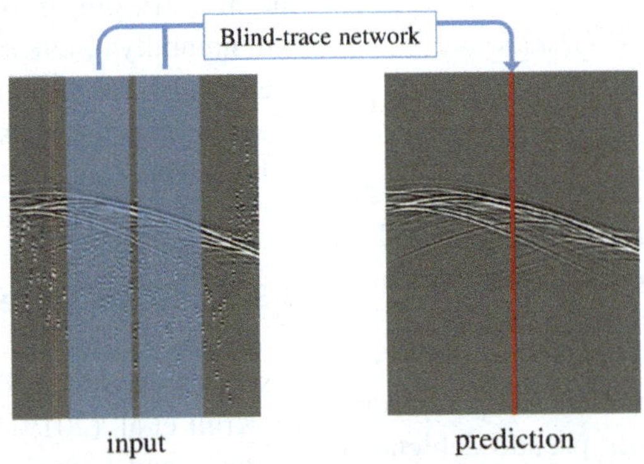

Fig. 3.10 The basic concept of the blind-trace constraint. The left image shows the input data (pseudo-deblended), while the right image depicts the predicted output (deblended data). To reconstruct the red-marked trace, the network utilizes only the surrounding blue-shaded area from the input, deliberately excluding the original noisy trace itself

convolutional neural network (BTN) for the self-supervised seismic deblending task, as illustrated in Fig. 3.11. The proposed BTN is constructed based on the U-Net (Ronneberger et al., 2015), with the framework unchanged, adapting all original computational components to a restrictive receptive fields version.

The process begins by duplicating the input seismic image and rotating it independently by 90 and 270°, shifting the left and right directions of the original common receiver gather upward, respectively. These two duplicated gathers are then independently fed into the network. The blind-trace network can be decomposed into two major part, namely the blind-trace U-Net and the aggregation block. Within the blind-trace U-Net, a significant constraint is strictly enforced to limit the receptive field of all computations "half-sided". Specifically, for every row in the input, the output row at the corresponding index is computed only based on its upward region (including itself to allow the implementation of skip connections in residual blocks He et al., 2016). Therefore, the network must leverage the coherent features for each trace independently from its left or right neighboring traces. This process is illustrated in Fig. 3.11. An example trace is marked in red while its left and right adjacent regions are highlighted as blue and green, delineating the informative areas utilized for prediction. For traces at the edges, only one side of their neighboring regions is valid to provide information and the other side is padded by zero values. After the blind-trace U-Net, the bottom row is cropped and another row of zero is padded at the top for each feature map to exclude the target trace from its receptive field. At this stage, the "blind-trace" constraint is finally achieved. The two output features are then rotated back to the original direction and concatenated. Two layers of 1 × 1 convolution can merge the single-sided features give the final prediction. Although only one trace is highlighted for illustration in Fig. 3.11, the entire image is processed in a single pass, and the network generates predictions for all traces concurrently.

With the aim of confining the informative region to the upward side for all traces throughout the blind-trace U-Net, we have modified all convolutional and down-sampling layers in the original U-Net, as illustrated in Fig. 3.12.

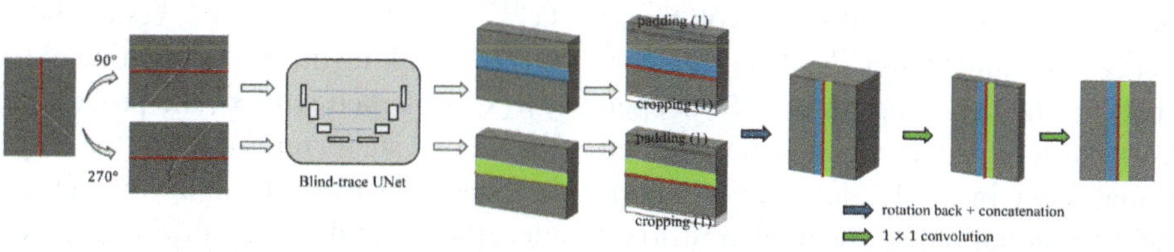

Fig. 3.11 Structure of the blind-trace network (BTN). While only one trace is marked as red for better delivering the idea, the blind-trace constraint is applied to all traces in the image simultaneously during a single forward pass. The patches in blue and green highlight the regions used to predict the deblended result for the marked trace. Initially, the target trace is included in the patches after the blind-trace U-Net. However, through subsequent padding and cropping layers, the red trace is excluded, making the network fully compliant with the blind-trace constraint

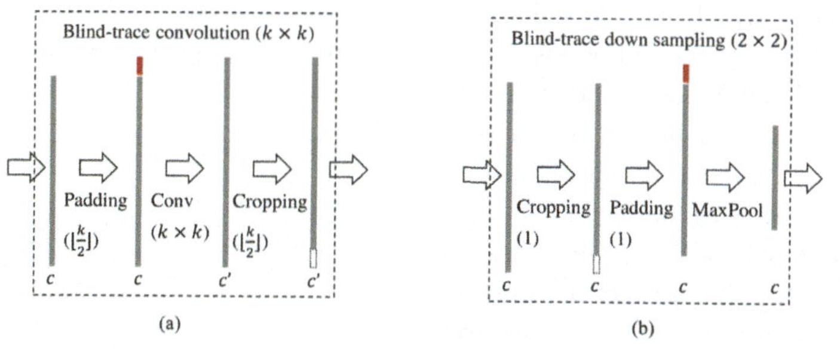

Fig. 3.12 Modified computations to incorporate the single-sided constraint in the blind-trace network (BTN). (**a**) Modified convolutional layer; (**b**) Modified max-pooling layer

(1) Convolution. Prior to each convolutional layer, zero-padding is applied to the feature maps. For a $(k \times k)$ convolutional layer, half the filter size ($\lfloor k/2 \rfloor$) rows filled with zeros are padded at the top of the feature maps, while the same number of $\lfloor k/2 \rfloor$ rows are cropped at the bottom following the convolution.

(2) Max-pooling. Before max-pooling, one line of zeros is padded at the top of the feature maps and one line is deprecated from the bottom. There is no need to modify the up-sampling layers, as the receptive field remains unaffected with the adjustment of the max-pooling layers.

Incorporating the described modifications ensures that every row of the predicted feature maps relies solely on itself and the region above it in the input, preventing any leakage of information from below. In our approach, the blind-trace U-Net has 4 encoding and their corresponding decoding blocks, each containing two concurrent and identical modified convolutional layers and a max-pooling (up-sampling for expanding blocks) layers. The channel number after the first layer is 32, and is doubled by the first convolutional layer in each encoding block, and halved correspondingly in the decoding blocks. To deepen the network structure, we replace the original convolutional layer in the U-Net with a "blind-trace" crafted residual block, meaning a residual block (He et al., 2016) where all convolutional layers are changed into the "blind-trace" version above. Batch Normalization layers in the original residual blocks are preserved in the decoder path. In the aggregation block, the concatenated feature map has 64 channels, which will be reduced to 32 and 1, respectively by the following 1×1 traditional convolutional layers. The nonlinear ReLU activation is utilized in the intermediate convolutional layers, while the leaky ReLU is applied in the final convolutional layer.

The proposed blind-trace network trained with the blending loss is a hybrid approach encompassing the advantages of both inversion-based and filter-based methods. The network's extensive parameters enable the implementation of advanced nonlinear filters that can effectively capture the highly abstract information and learn the coherency for seismic data. Simultaneously, in challenging cases when coherent signal is severely contaminated by strong noise and the filters fail to leverage coherency, the blending loss ensures the reconstruction of signals through the inversion constraint. This single-stage deblending framework is straightforward and eliminates the need for exhaustive hyperparameter tuning.

3.3.3 Two-Stage Amplitude Refinement with BTN

The deblending performance of a single-stage algorithm adopting the direct prediction from the blind-trace network may suffer when coherent signals are entirely obscured by potent blending noises. To address this issue, we propose a two-stage workflow as depicted in Fig. 3.13. After completing n epochs in the training stage using Eq. (3.11), we collect the predicted deblending outcome by the converged model in the first training stage denoted as $\mathbf{x}' = f_{\theta^*}(\mathbf{x}) = \{x'^{(1)}, \ldots, x'^{(N)}\}$, with θ^* being the converged weights. Taking the new prediction of the network $f_{\theta^*}(\mathbf{x}')$ obtained with $\mathbf{x}'$, the process proceeds to the second stage, referred to as the amplitude tuning stage. Instead of tuning the network parameter, the input data initialized as $\mathbf{x}'$ is iteratively updated using an updated loss function, which is defined as the weighted sum of the blending loss and the identity loss as

$$\mathcal{L}_{\mathbf{x}'}(\mathbf{x}, \mathbf{x}', f_\theta(\mathbf{x}'))=w_1\mathcal{L}_\mathrm{b} + w_2\mathcal{L}_\mathrm{I}, \quad (3.12)$$

where $\mathcal{L}_\mathrm{b}$ is the original blending loss

$$\mathcal{L}_\mathrm{b}(\mathbf{x}, f_\theta(\mathbf{x}')) = \frac{1}{N}\sum_{k=1}^{N} ||x^{(k)} - \mathcal{F}^{-1}(\Gamma^H\Gamma\mathcal{F}(f_\theta(x'^{(k)})))||_1, \quad (3.13)$$

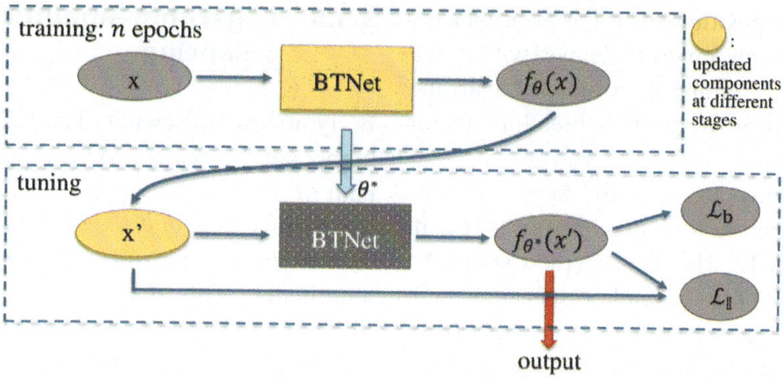

Fig. 3.13 The two-stage amplitude refinement algorithm. The top and the bottom rows represent the first training stage and the bottom amplitude tuning stage, respectively. The yellow colored BTN denotes the trainable part during the first stage. Once the training is converged, all weights θ^* are set unchanged and carried over to the second stage. In the tuning stage, the predictions from the training stage $\mathbf{x}' = f_{\theta^*}(\mathbf{x})$, colored by yellow, is adopted as the initial input. Instead of training the network, the input image is updated as the tuning loss in Eq. (3.12) is minimized

and $\mathcal{L}_{\mathbb{I}}$ is an identity loss defined as

$$\mathcal{L}_{\mathbb{I}}(\mathbf{x}', f_\theta(\mathbf{x}')) = \frac{1}{N}\sum_{k=1}^{N} ||x'^{(k)} - f_\theta(x'^{(k)})||_1. \tag{3.14}$$

Under the ideal condition, the $\mathbf{x}'$ and $f_{\theta^*}(\mathbf{x}')$ would be identical if no residual noise remains in the prediction after the first stage, meaning that both the initial input and prediction are clean. On the other hand, if there were any remaining noise in $\mathbf{x}'$, it will be further removed by the BTN in the second stage as well. As the tuning stage proceed, both the input and output are expected to be deblended data. The identity loss plays a crucial role to achieve reflect this assumption. More importantly, while the BTN is capable of strictly blocking the leakage of noises from the input to the prediction, it can also prevent the noise from the prediction from leaking into the input reciprocally. Nevertheless, the leakage might still transpire from neighboring areas in the prediction, returning to the input, which is not stoppable by the BTN. Specifically, while the noise in the target input trace is invisible itself, it is still available for the adjacent traces and could possibly exist in their predictions. This could lead to its propagation back to the updated input during the tuning stage. To prevent such propagation and unwanted artifacts, $\mathcal{L}_{\mathbb{I}}$ stringently regulates the inversion by enforcing the stationary of clean signal in both the input and output when processed by the blind-trace network.

3.3.4 Blended Acquisition Strategy

To highlight the effectiveness of the proposed method, we investigate three separate blending schemes.

(1) 'alternate' blending. With this blending approach, two successive sources are fired with a relatively short and random dithering time interval. As a result of the short-time dithering, the amplitudes of both signals and noises are at comparable levels. Our experiments first involve the Sigsbee dataset in which only a few dominant reflections are presented as presented in Fig. 3.15a. We then employ the SEAMII-unconventional model (Regone et al., 2017) to simulate a more complex dataset illustrated in Fig. 3.16a. In the SEAMII dataset, we collect 200 unblended common shot gathers with dimensions of (1600×751) along the receiver and time axis, respectively. The receivers are uniformly distributed with a 6.25 m distance, and the sampling ratio is 6 ms. The

shot gathers positioned on the even indexes are added with random delays within the range of (0, 1] s and then blended with the odd ones to accomplish a blending factor of 2.

(2) 'half' blending. In the 'half' blending strategy, the dithering time is relatively longer in the range of $[2, 3]s$. This method shifts the second half of the shot gathers and adds them to the first half. Consequently as depicted in Fig. 3.17a, severe interference happens in the bottom left and upper right corners in the input images. This is because the blending noise in the bottom left corner is the early arrivals from the second half, presenting much higher amplitudes than the signals in late arrival from the first half traces. The signals in the bottom left region can be completely obscureddue to the high amplitude contrast. Again, the SEAMII-unconventional model is used to synthesize the unblinded common shot gathers. The same 200 individual sources are deployed at the surface of the SEAMII-unconventional model with 1600 receivers and 750 time samples are collected in each trace with a sampling rate of 6 ms. Shots indexed ranging from 101 to 200 in the second half are blended accordingly to the first half of shots 1 to 100.

(3) 'continuous' blending. In this configuration, the simulate the simultaneous sources of streamer acquisition is simulated. The BP2004 velocity model serves as the basis for generating 200 continuous unblended shots, which are recorded at intervals of 5 to 6 seconds. We deploy 959 receivers along the drifting cable and collect 3334 time samples for each trace. The sampling rate is 6 ms such that the total sampling time is $T \approx 20s$). In this dataset, the number over interfered sources (blending factor) may reach up to 5, which indicates an aggravated noise level compared to the other two blending strategies.

3.3.5 Logarithm Amplitude Balancing

Early arrivals in seismic data tend to have much higher amplitudes due to the proximity and direct path of the seismic waves, whereas late arrivals, which travel longer distances and encounter more subsurface complexities, typically have lower amplitudes. This disparity can complicate the training and leads to over-weighted optimization for early arrivals. Therefore, an amplitude balancing method is essential as a data pre-processing step to influence the effectiveness of the deblending process.

We use a log1p operation for balancing the amplitudes of the blended data in order to properly attenuate early arrivals and amplify late arrivals. Because the deblending process requires to preserve the true amplitude of the seismic record, the pre-process approach must be reversible and the prediction should be scaled back by log1p^{-1} prior to computing the predicted pseudo-deblended data, and re-balanced to align the amplitude range of the input for loss computation in Eq. (3.11). The log1p and log1p^{-1} operators are defined as

$$\text{log1p}(x) = \text{sgn}(x) \log_{10}(|sx| + 1), \tag{3.15}$$

and

$$\text{log1p}^{-1}(x) = \frac{1}{s}\text{sgn}(x)(10^{|x|} - 1). \tag{3.16}$$

As depicted in Fig. 3.14, the log1p operator effectively enhances the late arrivals while suitably adjusting the input amplitudes into a reasonable range.

3.3.6 Results of One-Stage Deblending

The blending and pseudo-deblending operations in Eq. (3.11) require extensive matrix multipli-

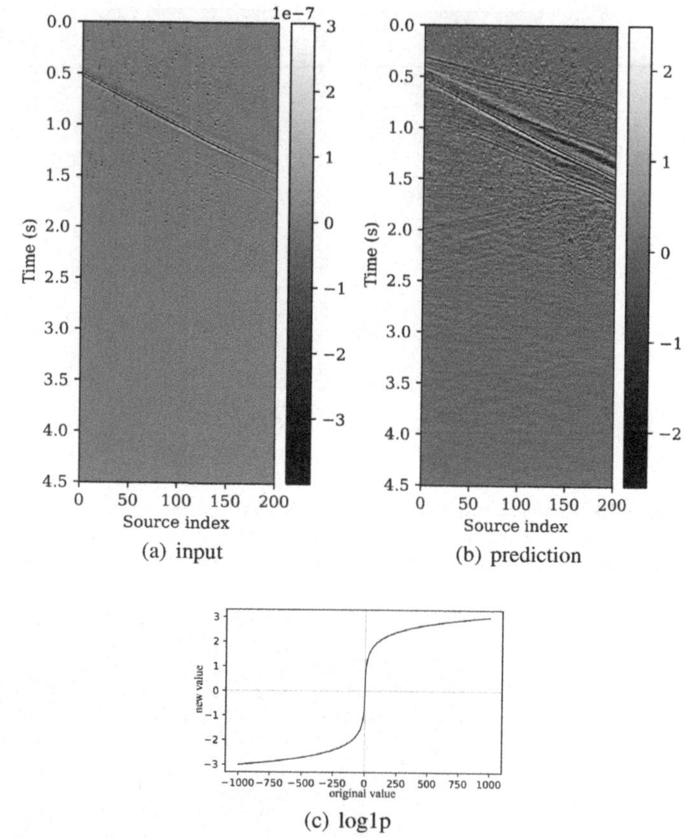

Fig. 3.14 Illustrate of the scaling operation on the SEAMII-unconventional dataset with 'alternate'-blended scheme. (**a**) Raw seismic data; (**b**) Scaled data by the proposed amplitude balancing method Eq. (3.15); (**c**) Plot for amplitude balancing function log1p in Eq. (3.15)

cations in the f-x domain, resulting in exhaustive computational and memory pressure for training. Therefore, we conduct these processes directly in the time-space domain without FFT. Specifically, the dithering time is directly adopted to shift the shot gathers along time axis, supported by cropping, padding and summation operations. During the training stage, the parameters of the blind-trace network are randomly initialized. And the network will be trained directly on the pseudo-deblended data till convergence. The direct predictions from BTN at the end of the training stage are presented in Figs. 3.15 to 3.19 and we compare the proposed approach with a conventional iterative thresholding method by Mahdad et al. (2011, 2012).

Figures 3.15 and 3.16 displays the retuls of the 'alternate' blending strategy. As shown in Fig. 3.15a, the focus is on reconstructing the dominant early events without balancing amplitudes using the log1p operator. It is evident that both the conventional iterative thresholding method and the proposed approach deliver satisfactory deblending results when dealing with simple and coherent patterns, as seen in Fig. 3.15c, d. However, when faced with more complicated signal patterns, as shown in Fig. 3.16, the conventional method struggles while the proposed method exhibits satisfactory deblending performance. The relative small dithering time between two neighboring shots avoids the mixture of early arrival and late arrivals, resulting in mitigated amplitude contrast issue. However, this results in noises scattered throughout the entire common receiver gather. Although the conventional method successfully recovers the strongest events in Fig. 3.16c, g, it fails to fully remove the noise. In contrast, the proposed BTN can effectively eliminates noise in both shallow and deep part as well as achieving accurate reconstruction for both low and high-frequency signals, as shown in Fig. 3.16d, h.

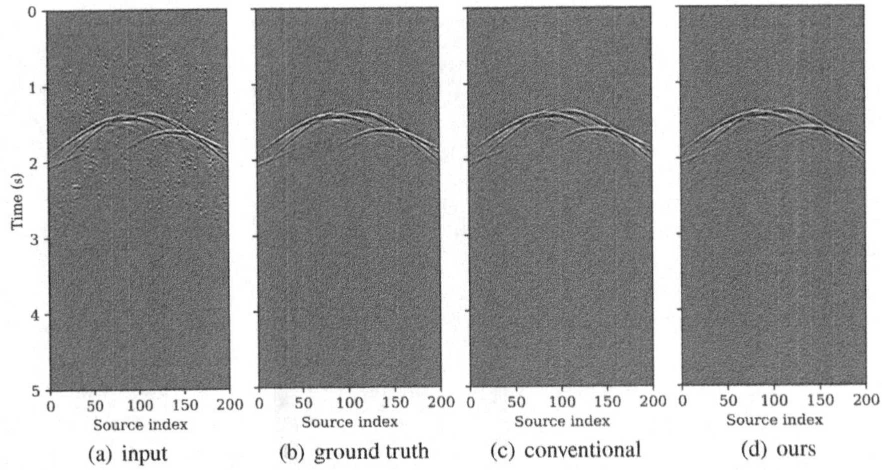

Fig. 3.15 Example result of the 'half'-blended Sigsbee data in the common receiver domain after the training stage. (**a**) and (**b**) The pseudo-deblended input and unblended ground truth, respectively. (**c**) and (**d**) The deblended prediction by conventional method and proposed BTN, respectively

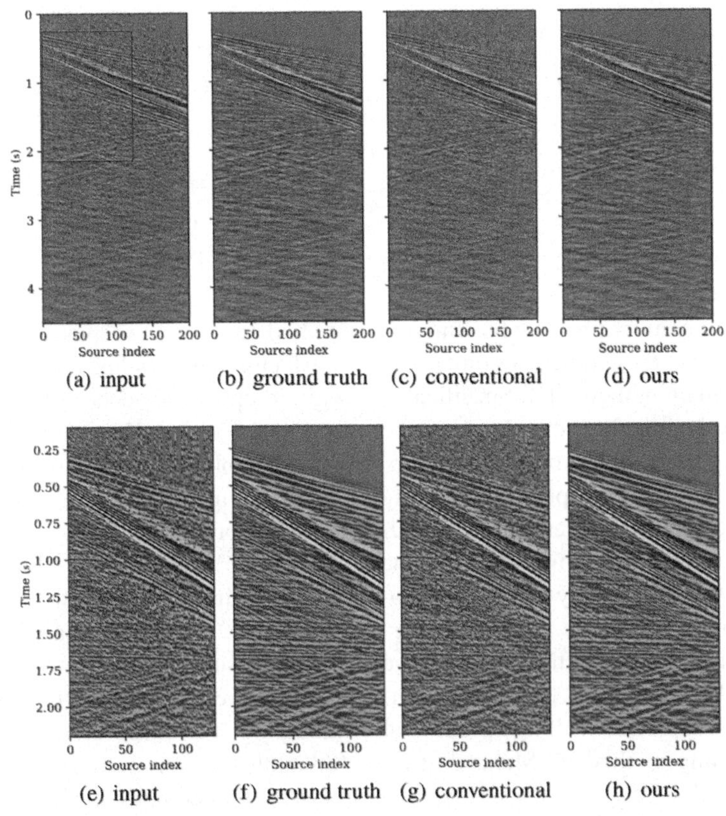

Fig. 3.16 Example result for the SEAMII-unconventional data under 'alternate' blending strategy in the common receiver domain after the training stage. (**a**) and (**b**) The pseudo-deblended input and unblended ground truth, respectively. (**c**) and (**d**) The deblended prediction by conventional method and proposed BTN, respectively; (**e**)–(**h**) Enlarged plots of the emphasized area in the red rectangular accordingly

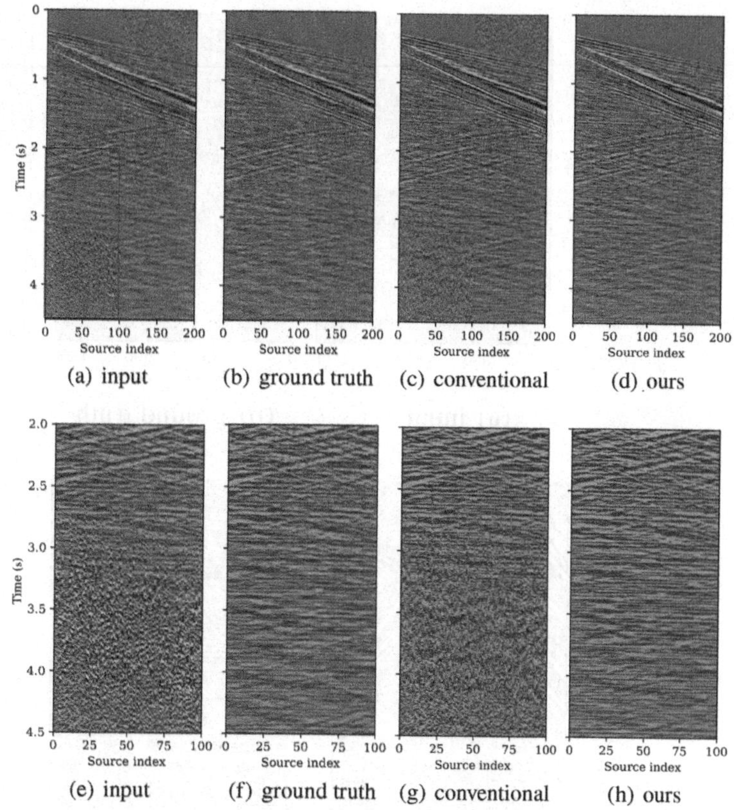

Fig. 3.17 Example result for the SEAMII-unconventional data under 'half' blending strategy in the common receiver domain after the training stage. (**a**) and (**b**) The pseudo-deblended input and unblended ground truth, respectively. (**c**) and (**d**) The deblended prediction by conventional method and proposed BTN, respectively; (**e**)–(**h**) Enlarged plots of the emphasized area in the red rectangular accordingly

The 'Half' blending strategy introduces further complexities for deblending algorithms. This scheme leads to concentrated noise by prolonged dithering time, and can significantly impacts the weaker amplitudes of late arrivals that are heavily masked by high-amplitude noises originating from early arrivals. Fig. 3.17a–d illustrate the deblending results and the highlighted parts are presented in Fig. 3.17e–h, offering a closer look at these effects. Conventional filter-based methods are prone to fail when coherent patterns are completely erased by the noises. Additionally, the problem of amplitude imbalance can considerably influence the deblending outcome. The conventional method performs deblending by iteratively estimating and eliminating blending interferences. As a result, even slight inaccuracies in noise estimation can lead to serious contamination of the hidden weak signals as depicted in Fig. 3.17g. In contrast, the coherent events buried under strong blending noises are precisely reconstructed in Fig. 3.17h. This indicates that the blind-trace network functions not only as a denoising filter, effectively separating clean signals from blending noise based on their coherency, but also as a solution to the under-determined inverse problem when coherent signals are completely buried.

Figures 3.18 and 3.19 display the common shot gathers of the deblended results for the same dataset in Figs. 3.16 and 3.17, respectively. It is evident that the proposed BTN method successfully removes the blending noise and reconstructs the separate sources accurately.

3.3.7 Results of Two-Stage Deblending

Due to the consecutive source firing, larger blending factors are obtained using the 'continuous' blending scheme. As depicted in Fig. 3.20a the highlighted section in Fig. 3.20b, the signals are obscured by several layers of blending noise,

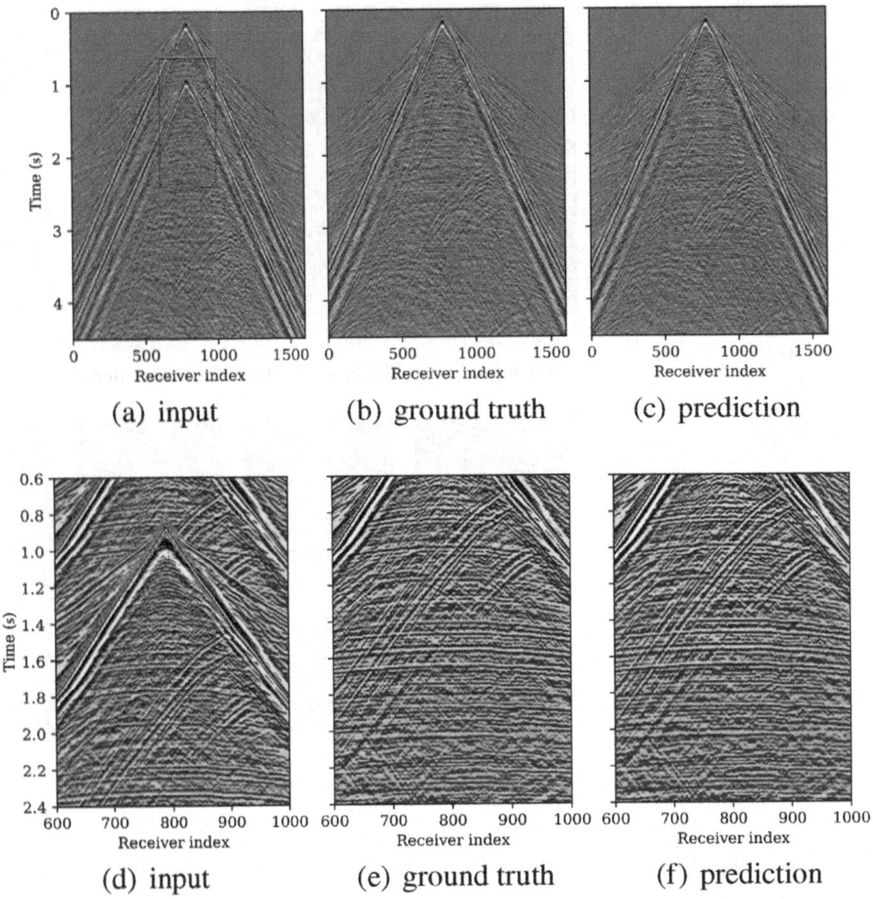

Fig. 3.18 Example result for the SEAMII-unconventional data under 'alternate' blending strategy in the common shot domain after the training stage. (**a**)–(**c**) The pseudo-deblended, unblended truth, and deblended prediction by proposed BTN, respectively; (**d**)–(**f**) The corresponding enlarged plots of the emphasized section in the red rectangular

demonstrating limited coherency from weak signals in the input gather compared to the ground truth in Fig. 3.20c. The one-stage result in Fig. 3.20d shows that artifacts exist and the reconstructed events are not accurate. To improve the deblending performance, the training-stage result (Fig. 3.20d) is collected after 200 epochs of training and passed to the second tuning stage as the initial guess for the input image. During the tuning stage, the input image is optimized with respect to the tuning loss Eq. (3.12). And the updated result is presented in Fig. 3.20e. After this process, many intricate and weak events that are primarily missing in Fig. 3.20d are restore. A detailed trace-wise comparison is presented in Fig. 3.21. When comparing Fig. 3.21b, c, it is evident that the predicted trace becomes more accurate by the tuning stage and more patterns are reconstructed. We also present an early stopping result for the tuning stage in Fig. 3.20f, where the network is only trained for 5 epochs in the first stage as apposed to adequately converged. After the same tuning iterations, the tuning result also effectively reconstructs the missing signâls in Fig. 3.20g. In this example, the amplitude balancing is reverted by $\log 1\mathrm{p}^{-1}$.

A quantitative comparison for the three blending schemes is summarized in Table 3.1. Significant improvement for all three blending schems are obtained through both training and tuning

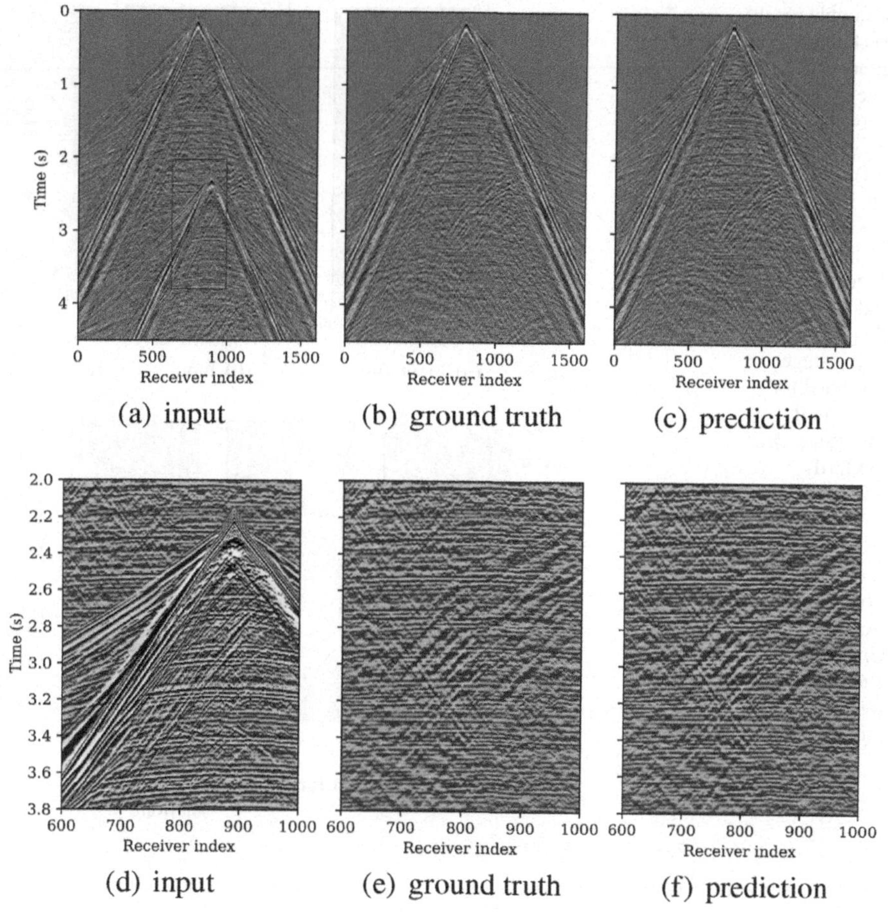

Fig. 3.19 Example result for the SEAMII-unconventional data under 'half' blending strategy in the common shot domain after the training stage. (**a**)–(**c**) The pseudo-deblended, unblended truth, and deblended prediction by proposed BTN, respectively; (**d**)–(**f**) The corresponding enlarged plots of the emphasized section in the red rectangular

stages, showcasing the effectiveness of the proposed approach.

3.4 Summary

In this chapter, we proposed two effective seismic data deblending methods to address network generalization issues and the lack of labeled datasets commonly encountered in deep neural network development.

First, we introduced a physics-augmented adaptive training workflow to develop a deep neural network for seismic deblending. The strength of our method lies in its innovative workflow that incorporating both velocity model building and simulation procedures. This self-augmented training data generation mechanism enhances the model's generalization capability without requiring extensive labeled datasets.

Furthermore, we proposed an advanced self-supervised learning method for deblending, which completely eliminates the need for labeled datasets. We introduced a blind-trace network (BTN) that can be directly optimized using the blending loss, facilitating successful self-supervised learning without additional constraints. This approach allows the network to be trained directly on the target dataset, eliminating the need for noisy-clean training data

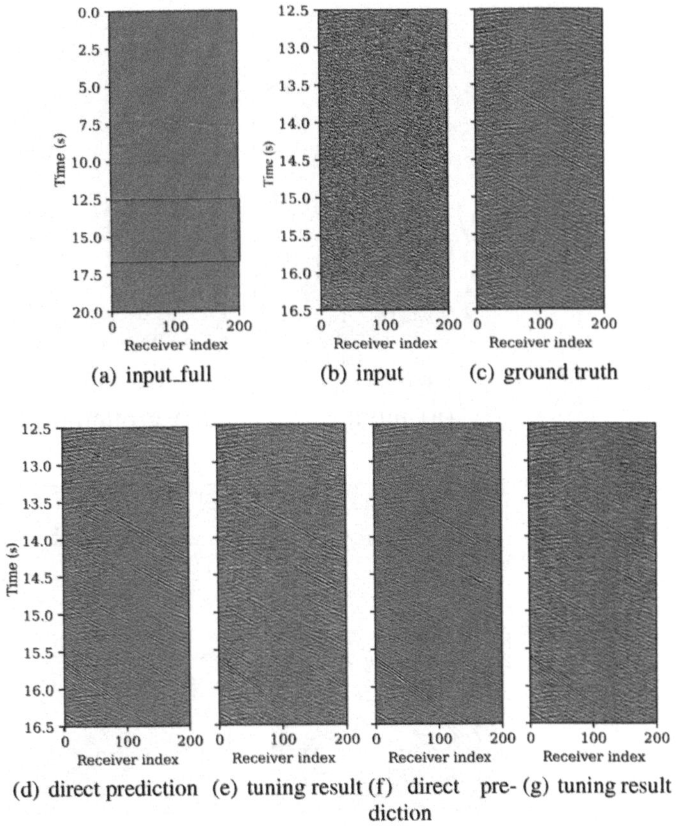

Fig. 3.20 Example result for the BP2004 dataset under 'continuous' blending strategy in the common receiver domain. (**a**) The entire seismic gather; (**b**)–(**d**) The enlarged plots for the highlighted section in the red box; (**b**) The pseudo-deblended data $\mathbf{x}$; (**c**) Unblended ground truth $\mathbf{x}^*$; (**d**) Direct prediction from the training stage (The BTN is trained for $n = 200$ epochs until adequate convergence.), i.e. $\mathbf{x}' = f_{\theta^*}(\mathbf{x})$; (**d**) Updated prediction after the tuning process using (**d**) as the initial input to the network; (**f**) Direct prediction from the training stage with only $n = 5$; (**g**) Updated prediction after the tuning process using (**g**) as the initial input to the network

pairs. The novel blind-trace constraint effectively addresses the under-determination issue present in conventional seismic deblending algorithms. While the BTN alone demonstrates exceptional deblending capabilities, we further proposed a two-stage seismic deblending workflow that includes an additional refinement step for the direct deblending predictions from BTN in challenging scenarios. The second stage, known as the tuning stage, combines conventional inversion concepts with state-of-the-art deep learning techniques to improve the accuracy of weak event reconstruction. An identity loss is introduced as a crucial regularization during this stage to ensure a non-trivial deblending solution.

Our proposed approach combines the benefits of both conventional filtering and inversion-based methodologies. This hybrid strategy allows the network to handle more complex scenarios while eliminating the need for exhaustive and time-consuming hyperparameter tuning in traditional methods. We explored three different simultaneous source acquisition strategies in our experimental setup, each presenting varied data complexity and noise levels, posing distinct challenges to the deblending algorithms. Nonetheless, the proposed method successfully addressed all scenarios, demonstrating remarkable performance across all datasets and proving its novelty and reliability.

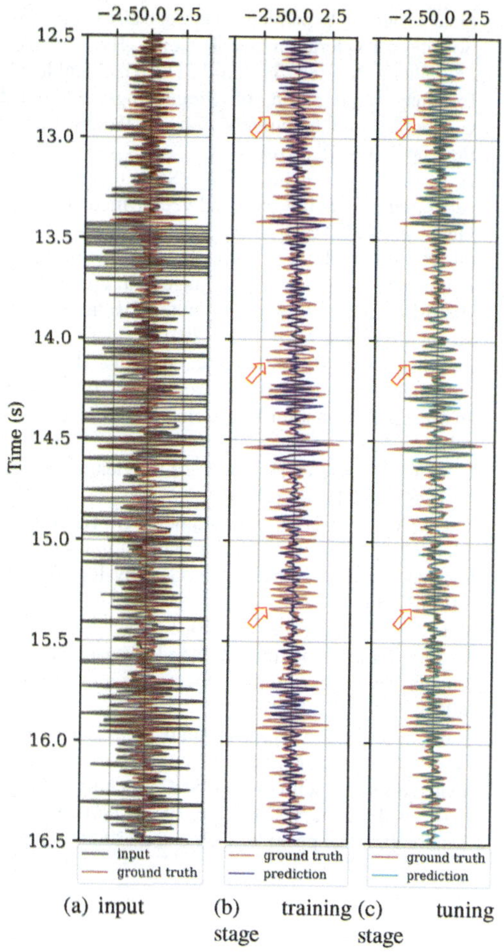

Fig. 3.21 Example of seismic traces in different stages for Fig. 3.20d, e, respectively. (**a**) The pseudo deblended input and the unblended ground truth; (**b**) The direct prediction from the first training stage in Fig. 3.20d; (**c**) The updated prediction from the second tuning stage in Fig. 3.20e

Table 3.1 Deblending loss/SNR(dB) comparison in different stages

Blending scheme	Pseudo-deblended	Training-stage	Tuning-stage
'alternate'	0.5178/1.42	0.1594/10.87	0.1141/13.51
'half'	0.0674/10.44	0.0169/26.02	0.0072/31.00
'continuous'	0.3250/-2.88	0.1176/5.20	0.0789/7.70

References

Abma, R., Howe, D., Foster, M., Ahmed, I., Tanis, M., Zhang, Q., Arogunmati, A., Alexander, G.: Independent simultaneous source acquisition and processing. Geophysics **80**(6), WD37–WD44 (2015)

Baardman, R.H., Hegge, R.F.: Machine learning approaches for use in deblending. Lead. Edge **39**(3), 188–194 (2020)

Bahia, B., Lin, R., Sacchi, M.D.: Iterative deblending with robust fourier thresholding. In: SEG Technical Program Expanded Abstracts 2020, pp. 3279–3283. Society of Exploration Geophysicists, Houston (2020)

Beasley, C.J., Chambers, R.E., Jiang, Z.: A new look at simultaneous sources. In: SEG Technical Program Expanded Abstracts 1998, pp. 133–135. Society of Exploration Geophysicists, Houston (1998)

Berkhout, A.E.: Changing the mindset in seismic data acquisition. Lead. Edge **27**(7), 924–938 (2008)

Chen, Y.: Deblending using a space-varying median filter. Explor. Geophys. **46**(4), 332–341 (2015)

Chen, Y., Fomel, S., Hu, J.: Iterative deblending of simultaneous-source seismic data using seislet-domain shaping regularization. Geophysics **79**(5), V179–V189 (2014)

Cheng, J., Sacchi, M.D.: Separation of simultaneous source data via iterative rank reduction. In: SEG Technical Program Expanded Abstracts 2013, pp. 88–93. Society of Exploration Geophysicists, Houston (2013)

He, K., Zhang, X., Ren, S., Sun, J.: Deep residual learning for image recognition. In: Proceedings of the IEEE Conference on Computer Vision and Pattern Recognition, pp. 770–778 (2016)

Huo, S., Luo, Y., Kelamis, P.G.: Simultaneous sources separation via multidirectional vector-median filtering. Geophysics **77**(4), V123–V131 (2012)

Krull, A., Buchholz, T.O., Jug, F.: Noise2void-learning denoising from single noisy images. In: Proceedings of the IEEE Conference on Computer Vision and Pattern Recognition, pp. 2129–2137 (2019)

Laine, S., Karras, T., Lehtinen, J., Aila, T.: High-quality self-supervised deep image denoising. In: Advances in Neural Information Processing Systems, pp. 6968–6978 (2019)

LeCun, Y., Bengio, Y., Hinton, G.: Deep learning. Nature **521**(7553), 436–444 (2015)

Mahdad, A., Doulgeris, P., Blacquiere, G.: Separation of blended data by iterative estimation and subtraction of blending interference noise. Geophysics **76**(3), Q9–Q17 (2011)

Mahdad, A., Doulgeris, P., Blacquière, G.: Iterative method for the separation of blended seismic data: discussion on the algorithmic aspects. Geophys. Prospect. **60**(4), 782–801 (2012)

Moore, I., Dragoset, B., Ommundsen, T., Wilson, D., Ward, C., Eke, D.: Simultaneous source separation using dithered sources. In: SEG Technical Program Expanded Abstracts 2008, pp. 2806–2810. Society of Exploration Geophysicists, Houston (2008)

Peng, C., Liu, B., Khalil, A., Poole, G.: Deblending of simulated simultaneous sources using an iterative approach-an experiment with variable-depth streamer data. In: 75th EAGE Conference & Exhibition Incorporating SPE EUROPEC 2013, pp. cp–348. EAGE Publications BV, Utrecht (2013)

Regone, C., Stefani, J., Wang, P., Gerea, C., Gonzalez, G., Oristaglio, M.: Geologic model building in SEAM Phase II—Land seismic challenges. Lead. Edge **36**(9), 738–749 (2017)

Richardson, A., Feller, C.: Seismic data denoising and deblending using deep learning. arXiv:1907.01497 (2019)

Ronneberger, O., Fischer, P., Brox, T.: U-net: convolutional networks for biomedical image segmentation. In: International Conference on Medical Image Computing and Computer-Assisted Intervention, pp. 234–241. Springer, Berlin (2015)

Wapenaar, K., van der Neut, J., Thorbecke, J.: Deblending by direct inversion. Geophysics **77**(3), A9–A12 (2012)

Zhang, Z., Wang, P.: Seismic interference noise attenuation based on sparse inversion. In: SEG Technical Program Expanded Abstracts 2015, pp. 4662–4666. Society of Exploration Geophysicists, Houston (2015)

Zu, S., Cao, J., Qu, S., Chen, Y.: Iterative deblending for simultaneous source data using the deep neural network. Geophysics **85**(2), V131–V141 (2020)

Zu, S., Zhou, H., Mao, W., Zhang, D., Li, C., Pan, X., Chen, Y.: Iterative deblending of simultaneous-source data using a coherency-pass shaping operator. Geophys. J. Int. **211**(1), 541–557 (2017)

4 Blind-Trace Network for Self-Supervised Seismic Data Interpolation

4.1 Introduction

4.1.1 Seismic Data Interpolation

Seismic data often exhibit non-uniform distribution and sampling. In the case of marine acquisitions, it is common for data to be densely sampled in one direction while sparse in another. This irregular spatial distribution typically results from cable drifting. Similarly, land-based seismic data collection often faces hurdles, including a significant number of defective traces and notable data gaps, which arise from various acquisition constraints. Direct application of standard seismic processing tools to such data can lead to strong aliasing and pronounced artifacts. This degradation in image quality can negatively impact amplitude versus offset (AVO) analysis, create imbalances in ray density for velocity tomography, and cause deviations in gradients for Full Waveform Inversion (FWI) updates. To address these issues, the data must be reconstructed into densely and uniformly spaced arrays.

Given the desired densely sampled (undecimated) shot gather symbolized as $\mathbf{y} \in \mathcal{R}^{n_r \times n_t}$ with the dimension of n_r traces and n_t time samples, respectively, we can denote the decimated observation as $\mathbf{x} \in \mathcal{R}^{n_r \times n_t}$ in which $\mathbf{x}$ has missing traces replaced by zeros vectors. Let $\mathbf{m}$ represent the set of indices of decimated traces in the observation, whose cardinality $|\mathbf{m}|_0$ indicates the count of decimated traces to be interpolated. Therefore, $r = |\mathbf{m}|_0/n_r$ defines the missing ratio of the shot gather. Based on this, we can formulate the relationship between $\mathbf{x}$ and $\mathbf{y}$ as

$$\mathbf{x} = \sigma_{\mathbf{m}}(\mathbf{y}), \quad \text{such that} \quad \mathbf{x}[:,i] = \begin{cases} 0 & \forall i \in \mathbf{m} \\ \mathbf{y}[:,i] & \text{else} \end{cases}, \tag{4.1}$$

where $\sigma_{\mathbf{m}}$ represents the masking process that set all missing traces indexed by i in $\mathbf{x}$ to zeros, mapping the undecimated gathers $\mathbf{y}$ to decimated ones $\mathbf{x}$. Consequently, our objective is to develop an interpolation function capable of converting the decimated observations into their complete form, i.e. $f_{\boldsymbol{\theta}} : \mathbf{x} \rightarrow \mathbf{y}$ with $f_{\boldsymbol{\theta}}$ the interpolation operator parameterized by $\boldsymbol{\theta}$.

4.1.2 History of Seismic Data Interpolation

There have been a variety of traditional approaches developed for seismic data interpolation, taking advantage of various assumptions and unique features of seismic events, such as coherency, low-rank nature, sparsity, etc. A prediction-error filter (PEF) in the frequency-spatial (f-x) domain is proposed by Spitz (1991) using the dependency of seismic traces. Oropeza and Sacchi (2011) leveraged the low-rank assumption of seismic data and introduced an

S. Wang et al., *Deep Learning for Seismic Data Enhancement and Representation*, Advances in Oil and Gas Exploration & Production, https://doi.org/10.1007/978-3-031-75745-7_4

iterative method utilizing multichannel singular spectrum analysis (MSSA) to reduce the rank of the seismic gathers and fill in the missing gaps. Zwartjes and Sacchi (2007) proposed interpolating irregularly sampled seismic data in the frequency-wavenumber (f-k) domain with a Fourier reconstruction method, addressing components in the data. A prevalent projection onto convex sets (POCS) algorithm, proposed by Abma and Kabir (2006), has been wildly adopted by conventional seismic data inteprolation algorithms. However, the POCS algorithm suffers from high computational costs with its iterative interpolation scheme and the difficulty in selecting an appropriate thresholding parameter value, which changes in each iteration. Another popular method, minimum weighted norm interpolation (MWNI), has shown promising results in previous methods but struggles with multiple cross-events in the presence of aliasing and is sensitive to the selection of the regularization term and parameters.

Compared to traditional interpolation methods, machine-learning approaches have been demonstrated to provide higher accuracy and reduced human interventions requirement. Jia and Ma (2017) interpolated the seismic data in patches using a support vector machine (SVM) and Chai et al. (2020a) employed a deep neural network to interpolate seismic gathers, with over a million patches involved for training. Chai et al. (2020b) also developed a 3D convolutional neural network and construct a 3D training dataset using both filed and synthetic seismic data. Mandelli et al. (2019) also employed patch-based training to jointly interpolate and suppress the noise in seismic data. Additionally, the Generative Adversarial Networks (GANs) are deployed by Chang et al. (2018, 2019) and Kaur et al. (2019, 2021) to generate complete seismic gathers conditioned on their decimated counterparts. End-to-end interpolation models are introduced by Wang et al. (2019) and Wang et al. (2020), by which Convolutional Neural Networks (CNNs) are trained with complete-missing pairs to learn the reconstruction of seismic traces. Pan et al. (2020) improved the reconstruction accuracy by introducing an improved partial convolution-based deep-learning network. Chai et al. (2020a) implemented a multiple-stage training framework to address the amplitude imbalance problem. Fang et al. (2021) introduced a texture loss term to enhance spatial structure continuity hence improving the interpolation performance. Despite the promising results achieved, all these methods are constructed with a supervised learning scheme, necessitating labeled seismic datasets and suffering from limited generalization ability to new datasets with different acquisition configurations and decimation patterns.

4.2 Self-Supervised Learning for Seismic Data Interpolation

4.2.1 History and Challenges for Self-Supervised Interpolation

The need for labeled datasets can be overcome by training a generative model with observed traces using a self-supervised learning scheme. A bidirectional long short-term memory (LSTM) was used by Yoon et al. (2020) to reconstruct the missing traces using their two neighboring traces. Sen et al. (2019) employed variational autoencoder (VAE), trained with a perceptual loss to learn the reconstruction of missing traces from random noises, but limited to regular sampling scenarios. Ulyanov et al. (2018) demonstrated that the deep neural networks for ill-posed problems (e.g. denoising and super-resolution) can be intrinsically regularized with prior information in the input data, without requiring explicit constraints. Building on this idea, Liu et al. (2021) introduced the deep seismic prior (DSP) method to learn the generation of coherent seismic events from Gaussian noises in a self-supervised learning manner, offering a more generic solution than Sen et al. (2019) and serving as the baseline for our experimental analysis. Kong et al. (2020) and Park et al. (2020) also adopted this concept and delivered promising results in their variant work of DSP. However, these techniques present challenges, including the sensitivity of interpolation

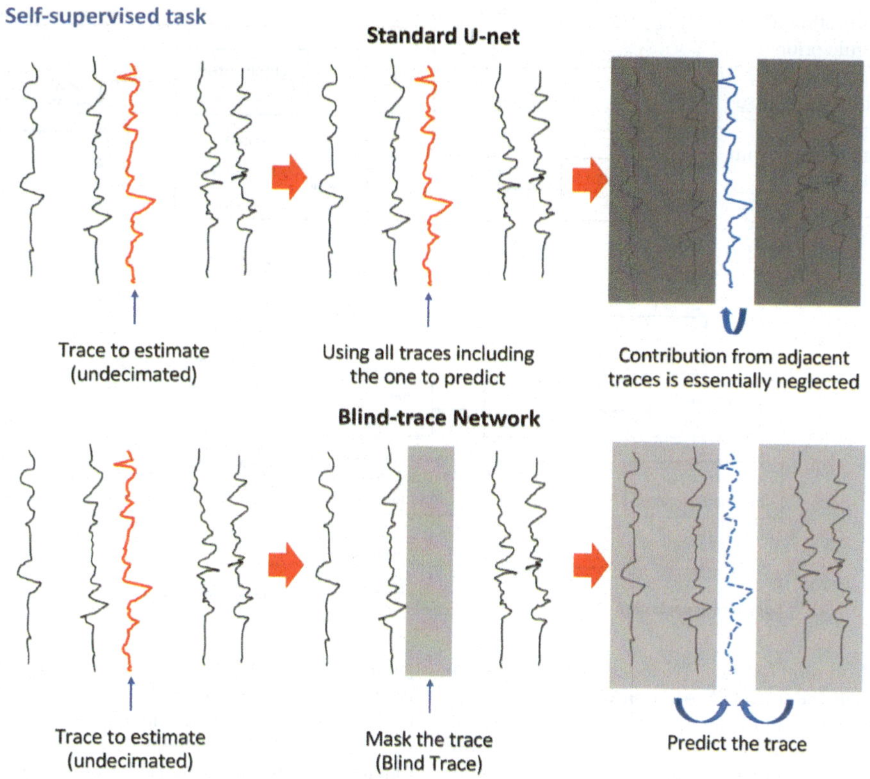

Fig. 4.1 Illustration of the principal distinction between a traditional CNN and the proposed Blind-Trace Network (BTN) for seismic data interpolation using a self-supervised learning scheme. In the BTN, the estimation of a target trace relies exclusively on its adjacent area, with the target trace itself excluded from the informative region. Conversely, a traditional CNN, without any constraints, tends to neglect adjacent traces, learning an identity mapping without effective interpolation functionality, thus resulting in a trivial solution

results to the initialization method when preparing the random noise input. They also lack the generalization capability to transfer learned seismic features across different seismic gathers due to varying initialization and the random relationship between the noise and the seismic features. Additionally, aliasing frequently complicates the interpolation process, posing significant obstacles to achieving accurate and reliable results.

Utilizing decimated seismic data as the network's input, as opposed to random noise, enhances efficiency by supplying the network with direct seismic priors, which in turn fosters more robust predictions. However, the application of a traditional CNN, such as the U-Net (Ronneberger et al., 2015), in such a self-supervised learning scheme can cause problems. In this scenario, the network can simply copy the input (including both known traces and missing traces set as zeros) instead of performing the intended interpolation. This issue, illustrated in the first row of Fig. 4.1, arises because the network tends to neglect the contribution from adjacent traces since the observed traces in the input can easily leak into the prediction. As a result, there's a risk of the network's prediction merely identical to the input, leaving the decimated traces unaltered. Consequently, effectively addressing this data leakage is crucial for the success of a self-supervised seismic data interpolation algorithm.

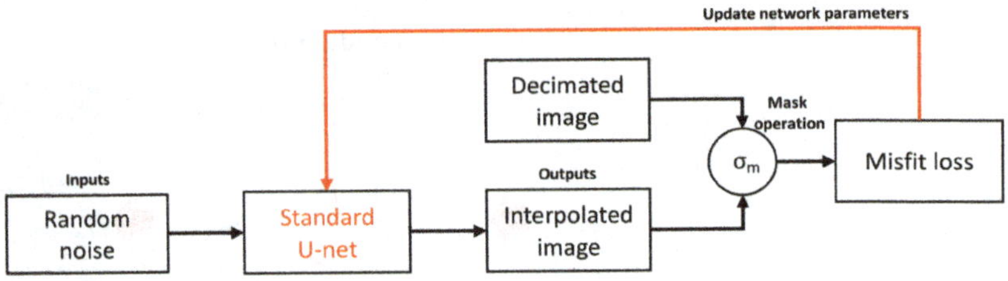

Fig. 4.2 Illustration for the deep seismic prior (DSP) algorithm. The network leans a generative method to produce complete shot gathers from random noise maps

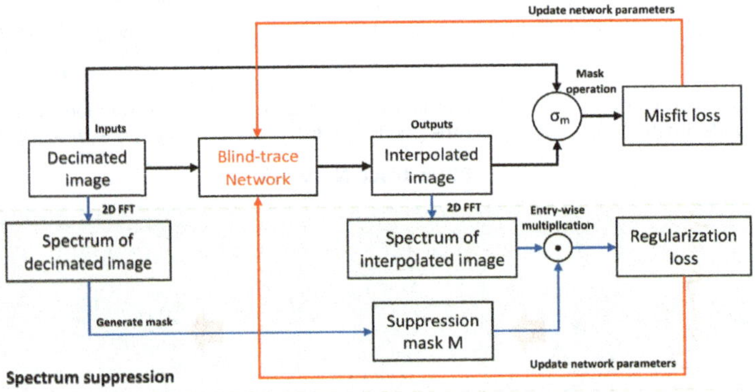

Fig. 4.3 Illustration for the BTN-based method. The upper half depicts the sole-BTN solution where the network is directly trained with the interpolation loss, whereas the bottom half adds additional spectrum suppression constraint to the training that can suppress the aliasing components in the frequency-wavenumber (f-k) domain, referred to as the BTN-SS method. For both alternatives, the network is directly provided with the decimated seismic gathers as input, replacing the traditional U-Net with the BTN. For the BTN-SS method, an f-k mask is generated by inferring a cone-shaped area in which the main energy is concentrated. The boundary of the cone shape is automatically determined by line detection algorithms. The frequency components outside the f-k mask are suppressed during training

4.2.2 Self-Supervised Seismic Data Interpolation with Blind-Trace Network

The self-supervised training scheme for seismic data interpolation is often accomplished by learning to reconstruct the undecimated traces, ignoring the decimated ones. The training loss is constructed as

$$L_\theta(\mathbf{m}) = \|\mathbf{x} - \sigma_{\mathbf{m}}(f_\theta(\mathbf{x}))\| \ , \qquad (4.2)$$

in which $\|\cdot\|$ denotes the l_1 norm. The network f_θ is optimized regarding the training loss and adopted for the inference phase after convergence. During the testing phase, the network is evaluated by the interpolation loss

$$L_\theta(\widetilde{\mathbf{m}}) = \|\sigma_{\widetilde{\mathbf{m}}}(\mathbf{y}) - \sigma_{\widetilde{\mathbf{m}}}(f_\theta(\mathbf{x}))\| \ . \qquad (4.3)$$

where $\widetilde{\mathbf{m}}$ is the complement of the set $\mathbf{m}$, representing the indices of undecimated traces that are not utilized for supervision during training. The complete gather $\mathbf{y}$ is only assumed available in the testing phase.

Generative interpolation methods, such as the DSP method shown in Fig. 4.2 involve learning a function that produces complete seismic gathers from random noises, i.e. $g_\theta : \mathbf{z} \rightarrow \mathbf{y}$, where $\mathbf{z}$ denotes the randomly generated noise vector or map that a complete image is conditioned on. Training these generative networks can be inefficient when working with multiple shot gathers and inherently adds variance to the interpolation results. To maximize the utilization of the observed traces and provide explicit priors for the network, we utilized the blind-trace network (BTN), enabling the replacement of random noises with decimated gathers while avoiding the trivial solution found in traditional CNNs. This workflow is illustrated in Fig. 4.3.

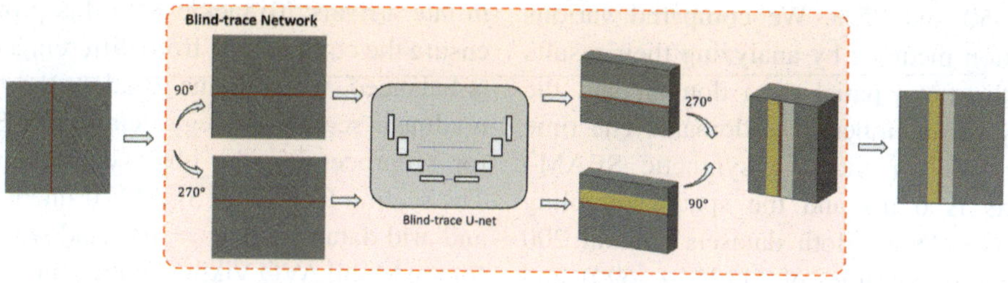

Fig. 4.4 Similar to Fig. 3.11. The blind-trace network is composed of two paths of blind-trace U-Net each constraining the receptive field for a target trace to its left and right neighbors, respectively. The resulting two feature maps are subsequently combined using (1 × 1) convolutional layers to generate the final prediction

Specifically, the BTN is effective in addressing the identity mapping issue when changing the input data from random noise to the decimated seismic gathers. As illustrated in Fig. 4.1, for each trace to be reconstructed, imposing the blind-trace constraint inherently conceals itself in from the informative region during the forward propagation. Amplitudes of each trace are predicted strictly and exclusively based on its neighboring traces, based on the assumption of continuity and predictivity between neighboring traces. Under this strict constraint, the network is essentially forced to learn interpolation, thereby preventing the identity mapping issue by design. The network structure is summarized in Fig. 4.4.

4.3 Blind-Trace Network for Irregular Seismic Data Interpolation

4.3.1 Data Preparation and Pre-Processing

In our experimental setup, we utilized two synthetic datasets, namely SEAM–II unconventional and SEAM–II arid, along with one field dataset (Mobil AVO Viking Graben line 12) to assess the proposed method. Both synthetic datasets are derived from the second SEAM project (Regone et al., 2017). The Barrett Unconventional model within this project is tailored to reflect the unique challenges encountered in the exploration and characterization of unconventional hydrocarbon reservoirs. It features a complex interplay of reservoir and overburden signals at varying amplitudes, coupled with geological noise from near-surface scattering. This model is characterized by numerous conflicting dips near the surface, which presents additional challenges in reconstructing early arrivals. The Arid model, on the other hand, is designed to mimic desert environments. It encompasses diverse geological features like karsts, wadis (buried river channels), sand and other unconsolidated sediments, outcropping bedrock, and sharp topography. This model is particularly challenging due to the abundance of weak diffractions, which are often incoherent and can increase the complexity of the data. The Mobile AVO Viking Graben line 12 dataset, a widely recognized marine dataset, is utilized to validate the performance of seismic data interpolation (Chai et al., 2020a,b) and inversion (Keys and Foster, 1998) algorithms. Unlike overly simplistic synthetic data, the SEAM datasets are industry-standard benchmarks that realistically simulate geological environments found in different regions globally. The Mobile dataset, being field data, provides an opportunity to assess the performance of each method under more complex, real-world conditions. We believe these datasets offer a comprehensive testing ground that accurately reflects the diverse challenges faced in seismic data analysis and interpolation.

In our study, we initially concentrated on assessing the effectiveness of the proposed interpolation method under random (irregular) decimation scenarios in Sect. 4.3.2, employing decimation ratios of 75 and 87.5%. Then we will extend the detailed discussion for regular decimation scenario in Sect. 4.4 with decimation ratios of 50 and 75%. We compared various interpolation methods by analyzing their results in both the time-spatial (t-x) domain and the frequency-wavenumber (f-k) domain. The time sampling rate for the two synthetic SEAM–II datasets is 6 ms and the spatial sampling distance is 6.25 m. Both datasets contain 200 shots, with dimensions of (1121 × 690) and (1600 × 751) for SEAM-II Arid and SEAM-II unconventional datasets, respectively. Here, 1121 and 1600 denote the number of receivers, while 690 and 751 represent length of the time axis. For the field data Mobil AVO Viking Graben line 12 dataset, there are 1001 shot gather images, each measuring (120 × 1500) with 120 receivers evenly spaced at 25 m intervals, and 1500 time samples at a 4 ms sample rate. All images' abscissas indicate the relative distance or time lapse from the first receiver location or recording time.

Although seismic data interpolation has been extensively studied, most approaches rely on a supervised learning scheme that requires complete shot gathers for network training. In this experiment, separate models are constructed to learn the reconstruction of undecimated traces for different datasets, and applied to their decimated ones. In this setup, the training dataset is always the same as the target dataset in our experiments. We demonstrate that this approach offers two key advantages: firstly, it enables the application of the model to all datasets, even in the absence of complete shot gathers, and secondly, it eliminates the need to address domain shift issues between datasets. Nonetheless, we still evaluated the models' generalization capabilities by splitting the shot gathers into training and testing set. We stress that the major distinction compared to the generative interpolation approaches is that the training shot gathers are all decimated seismic gathers.

The amplitude contrast is also a major issue in the seismic data interpolation task. When the amplitudes of early arrivals can be hundreds of times larger than that of late arrivals in raw seismic records, this amplitude imbalance may lead to a loss that is disproportionately influenced by high-amplitude early arrivals, thereby challenging the reconstruction process for low-amplitude events in late arrivals. To tackle with this problem and ensure the contribution from different amplitudes is balanced, we continue to adopt the logarithm nonlinear scaling strategy detailed in Eq. (3.15) for preprocessing the observed seismic traces. The scaling factors for SEAM–II unconventional and arid datasets are $s = 10^9$, and we set $s = 1$ for the Mobil AVO Viking Graben line 12 dataset. This invertible pre-processing operation resembles converting the original amplitudes into the decibel scale. To ensure enhanced visual quality and detailed presentation of the data, all seismic images in this chapter are displayed using this scaling method, with a 99% percentile clipping applied consistently. In evaluating the effectiveness of our interpolation methods, we employ various metrics, including mean absolute error (MAE), mean squared error (MSE), and signal-to-noise ratio (SNR). The SNR is calculated by

$$\mathrm{SNR} = 10\log_{10}\left(\frac{\|\mathbf{y}\|_2}{\|\mathbf{y} - f_\theta(\mathbf{x})\|_2}\right), \tag{4.4}$$

in which $\|\cdot\|_2$ denotes the l_2 norm, and $f_\theta(\mathbf{x})$ is the interpolated prediction from the network.

4.3.2 Experimental Results for Irregular Seismic Data Interpolation

In this section, we present the interpolation results for irregularly sampled seismic data with the SEAM–II arid dataset and the Mobil AVO Viking Graben Line 12 dataset. For each dataset, all seismic gathers are subjected to the same decimation ratio, but with varied random decimation patterns. These patterns reflect the non-uniform sampling typical of seismic data, characterized by missing traces or gaps occurring at different locations. It's noteworthy that the DSP method (Liu et al., 2021) and other generative methods with self-supervised learning have predominantly been designed to interpolate a single gather with a fixed noise input. Potential reasons for this include: (1) training a distinct model for each decimated gather is both time and computational expensive (2) a general model shared by the whole decimated dataset may yield suboptimal results since these models often rely on distinct random noise inputs, and (3) the preservation of input random noises is crucial, as even minor variations can result in significant differences in the reconstructed seismic data, and vice versa. To optimize resource usage, we have implemented the DSP approach using a shared network. For this, we pre-sample 200 different random noises as inputs and train the same network with all samples simultaneously. A traditional U-Net is constructed with an encoder of 4 blocks, and their corresponding blocks in the decoder, similar to the BTN. The feature map channels for both the U-Net and BTN are set as 32 in the first encoder block and doubled in the first convolutional layer of each contracting block). Networks for both the baseline method and our proposed approach are trained using the Adam optimizer, with a uniform learning rate of 0.001 for all experiments.

Figure 4.5 displays interpolation results of the SEAMII-arid dataset under the 75% random missing ratio. The top row displays the decimated data alongside the complete dataset for reference. Figure 4.5a, b depict the entire decimated and complete images, respectively. The subsequent second and third rows present the interpolated results and the corresponding residuals from the DSP method and our proposed BTN method, respectively. A notable observation from these results is that the DSP method's interpolation yields higher residuals throughout the entire seismic gather. In contrast, the BTN method achieves significantly lower residuals, predominantly concentrated in the near-offset traces. The scenario with an 87.5% random decimation ratio is similarly illustrated in Fig. 4.6 in the same format as Fig. 4.5. In this case, our proposed method maintains lower residuals and demonstrates robustness even with the increased decimation ratio. Conversely, the DSP method exhibits substantially higher errors that extend across the entire image. Furthermore, a comparative analysis in the frequency-wavenumber (f-k) domain is presented in Fig. 4.7. This comparison clearly reveals that our BTN method markedly outperforms the DSP approach in accurately reconstructing the f-k spectrum, underscoring the efficacy of our proposed technique in seismic data interpolation.

Table 4.1 provides a quantitative comparison for the randomly decimated SEAM-II arid dataset. As expected, the total error escalates with an increase in the number of decimated traces. Nonetheless, our proposed BTN method demonstrates superior performance over the DSP method across all metrics, in both 75% or 87.5% decimation scenarios. A significant advantage of the BTN method is its faster convergence compared to the DSP method, attributable to its initial interpolation input being decimated data rather than random noises. When running on the same RTX 8000 GPU, and using identical optimizers and learning rates, the DSP method requires 800 epochs to converge, whereas the BTN method achieves convergence in just 200 epochs. Although a single epoch takes longer to complete in the BTN method, the overall training duration is considerably shortened. This results in a time-saving of about 66%. Figure 4.8 presents the variation of MAEs during training. The training losses computed based on the undecimated taces are depicted by solid curves, while the testing/interpolation losses are delineated by the dashed ones. To be noted, the testing loss is only monitored for performance assessment, and is not used for weight updating or model selection during training. This figure clearly shows that the BTN method achieves a faster convergence speed and delivers quantitatively better interpolation results compared to the DSP method under various irregular decimation scenarios. In terms of training time, while the DSP method requires 27 hours on a single 48 GB Nvidia RTX 8000, our BTN method completes training in just 9.5 hours, leading to a time reduction of over 60%. When multiple GPUs are available, the training process can be further accelerated by distributed training, offering significant advantages in terms of resource utilization and computational efficiency.

Figure 4.9 presents the histogram of interpolation errors for the multi-shot training experiment using the SEAM-II arid dataset. From this figure, it's evident that the BTN method con-

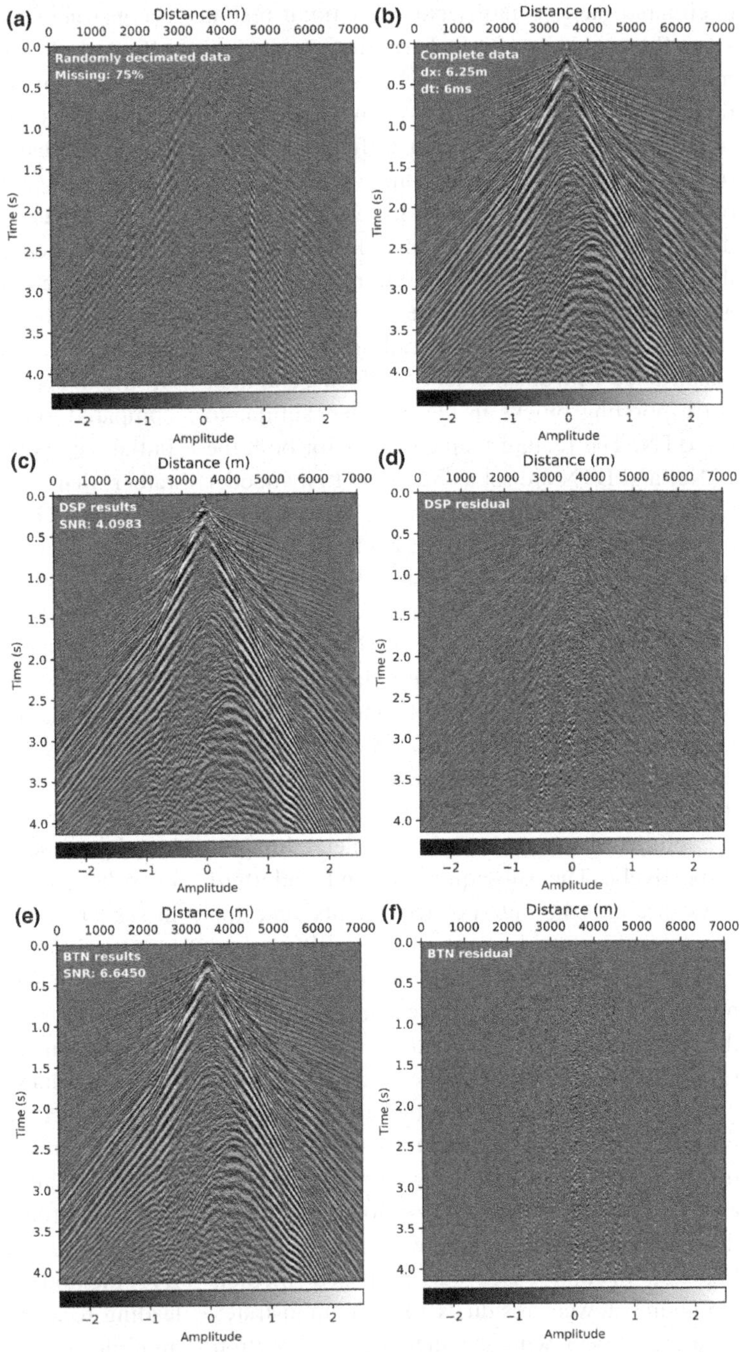

Fig. 4.5 Comparison for the randomly decimated SEAM–II arid dataset with a missing ratio of 75%. (**a**) Decimated observation where the missing traces are inserted as 0s; (**b**) Complete seismic gathers; (**c**) Interpolated result by the DSP method; (**d**) Difference between (**b**) and (**c**); (**e**) Interpolated result by the BTN method; (**f**) Difference between (**b**) and (**e**)

sistently results in smaller errors compared to the DSP method. Specifically, the most frequent errors (modes) for the BTN method are close to 0 under a 75% decimation ratio and around 0.1 under an 87.5% decimation ratio. This indicates a more accurate interpolation by the BTN method across different decimation scenarios. Figure 4.10 further elucidates the relationship between interpolation error and the size of the gaps in the seismic gather, where gap size for a missing trace is calculated by measuring the distance from itself to its closest undecimated traces. As expected,

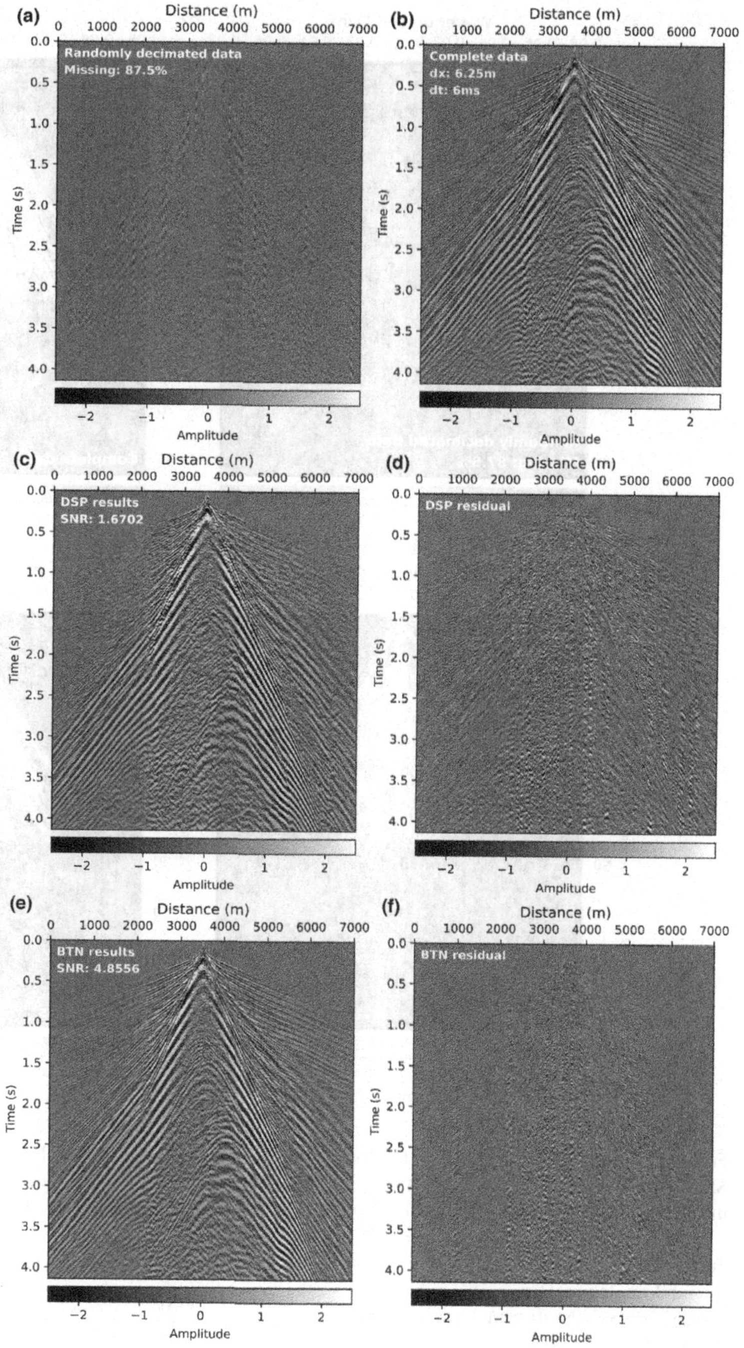

Fig. 4.6 Comparison for the randomly decimated SEAM–II arid dataset with a missing ratio of 87.5%. (**a**) Decimated observation where the missing traces are inserted as 0s; (**b**) Complete seismic gathers; (**c**) Interpolated result by the DSP method; (**d**) Difference between (**b**) and (**c**); (**e**) Interpolated result by the BTN method; (**f**) Difference between (**b**) and (**e**)

the Mean Absolute Error (MAE) tends to increase with larger gaps, reflecting the inherent challenge of interpolating over wider missing areas. The BTN method demonstrates robust performance and consistently outperforms the DSP method, showcasing its reliability and effectiveness in maintaining interpolation accuracy across various gap sizes in the seismic data.

The quantitative comparison for the Mobil AVO Viking Graben line 12 dataset is detailed in Table 4.2. For this dataset, we introduced an additional baseline method, training a standard U-

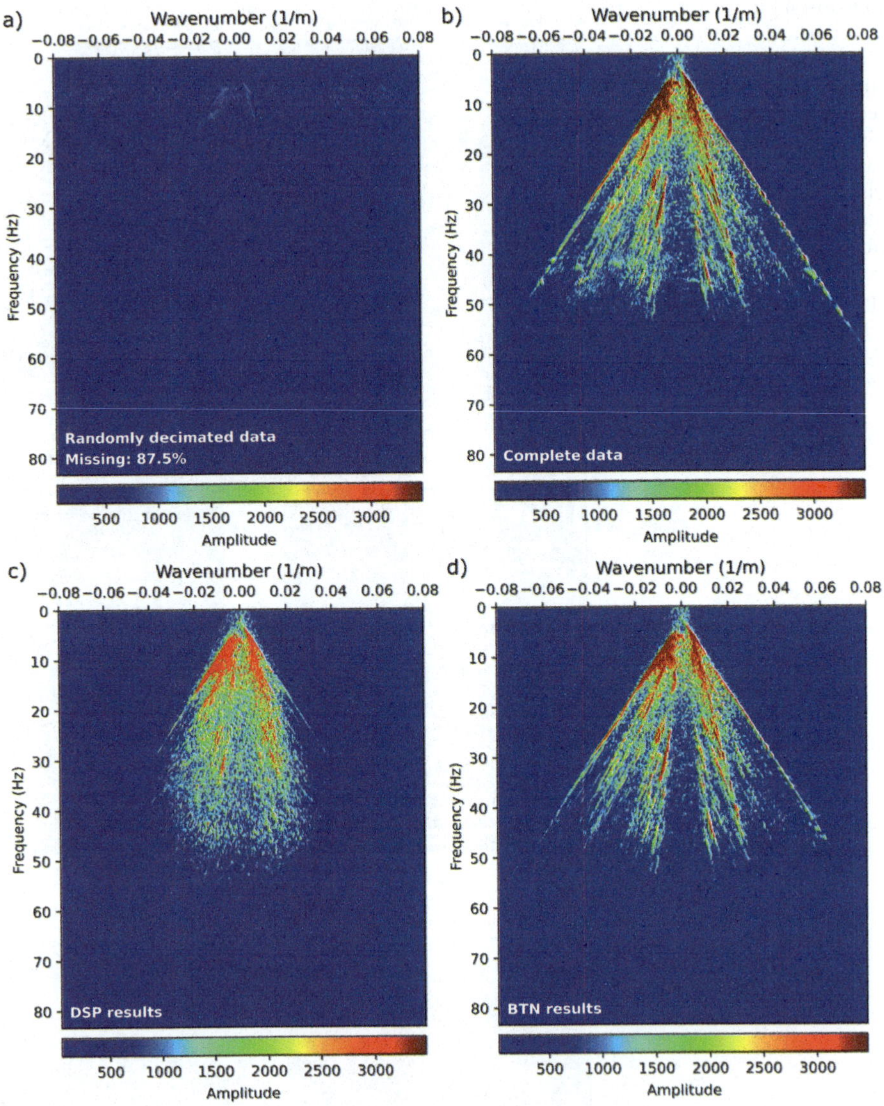

Fig. 4.7 Comparison for the randomly decimated SEAM–II arid dataset with a missing ratio of 87.5% in the frequency-wavenumber (f-k) domain. (**a**) Decimated observation where the missing traces are inserted as 0s; (**b**) Complete seismic gathers; (**c**) Interpolated result by the DSP method; (**d**) Interpolated result by the BTN method

Table 4.1 Summary of the random decimation scenario on SEAM–II arid dataset

Methods	Missing ratio	MAE	MSE	SNR	Epochs
DSP	75%	0.3460	0.4034	4.4982	800
BTN	75%	**0.1938**	**0.2154**	**7.2346**	200
DSP	87.5%	0.5071	0.8067	1.4818	800
BTN	87.5%	**0.3088**	**0.3843**	**4.7152**	200

The bold values highlights the best results

net with decimated gathers as input using a self-supervised learning scheme. Consistent with the findings from the synthetic data experiment, the proposed BTN method consistently outperforms this baseline, whereas the U-net approach yields the least favorable results. Figure 4.11 depicts the training loss variations of the three methods under the two designated missing rates. It is

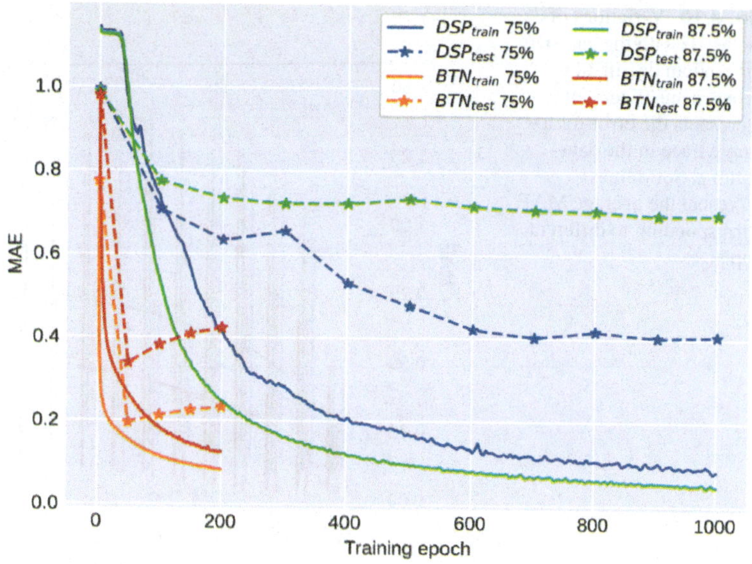

Fig. 4.8 Loss curve for randomly decimated SEAM–II arid data

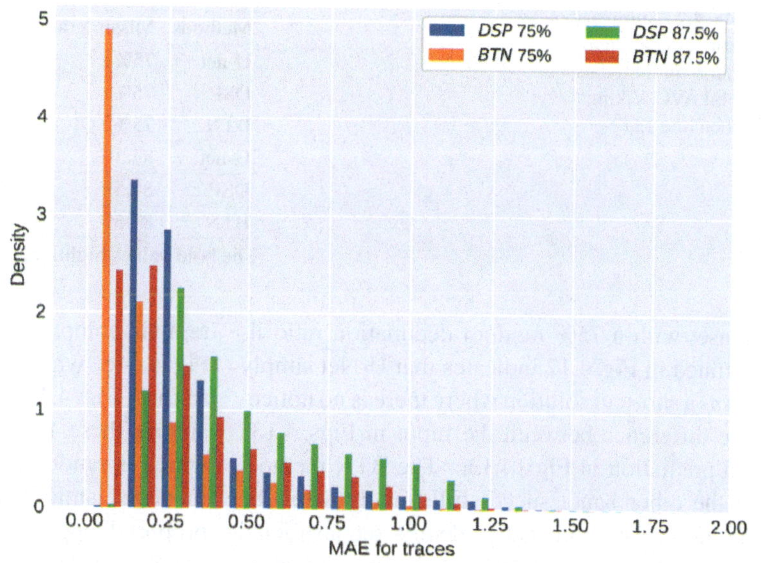

Fig. 4.9 MAE histogram for randomly decimated SEAM–II arid data

evident to find when the self-supervised learning scheme is performed on a traditional CNN, such as the U-net, the training loss can quickly converge to zero loss within a few epochs, but the corresponding testing loss is significantly high, indicative of a severe overfitting issue due to the identity mapping problem. This can be proved by Figs. 4.12, 4.13, 4.14, and 4.15 where the conventional U-net method essentially replicates the decimated input in the output, failing to reconstruct the missing traces. While the DSP method takes about 500 epochs to converge, the BTN method requires only 200 epochs. Despite this quicker convergence, the BTN method yields considerably lower testing errors compared to the DSP method. As the decimation ratio increases, the challenge of reconstructing decimated traces with seismic prior knowledge becomes more pronounced, as highlighted by the rising test loss for the DSP method in Figs. 4.14 and 4.15.

The qualitative comparison for the random decimated Mobil AVO Viking Graben line 12

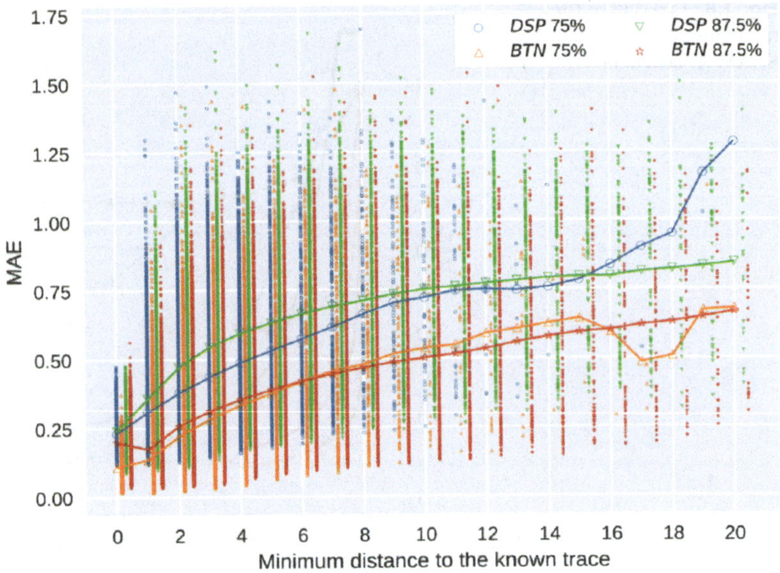

Fig. 4.10 Variations of the MAE with the gap size for random decimation experiments. Each dot represents the error for a single trace in the data volume. Solid lines represent the average MAE corresponding to different gap sizes

Table 4.2 Summary of the random decimation scenario on field dataset (Mobil AVO Viking Graben line 12)

Methods	Missing ratio	MAE	MSE	SNR	Epochs
U-net	75%	0.3743	0.3744	1.2465	<100
DSP	75%	0.3185	0.2778	2.6370	500
BTN	75%	**0.2041**	**0.1240**	**6.1044**	200
U-net	87.5%	0.4354	0.4370	0.5745	<100
DSP	87.5%	0.4186	0.5341	−0.2915	500
BTN	87.5%	**0.2799**	**0.2412**	**3.2056**	200

The bold values highlights the best results

dataset with a 75% random decimation ratio illustrated in Fig. 4.12 indicates that U-Net simply learns a shortcut solution where there is no noticeable difference between the input in Figs. 4.13a and prediction in Figs. 4.13c. The BTN method, on the other hand, successfully interpolates the missing traces with better details reconstructed compared to the DSP method in Figs. 4.13d, e. Figs 4.14 and 4.15 presents the comparison under a higher 87.5% random decimation ratio, following the same format as in Fig. 4.12. In these examples, particularly in Figs. 4.14d and 4.15d, noticeable artifacts and disconnected events, especially in early arrivals with large slopes, lead to aliased patterns. However, the BTN method displays a higher accuracy in reconstructing such challenging events. The f-k spectra of the first row in Fig. 4.14 is shown in Fig. 4.16. The spectra reconstructed by the two reference methods in Fig. 4.16c, d present noticeable artifacts in most regions compared to the BTN method result in Fig. 4.14e, which closely resembles the ground truth in Fig. 4.14b. A trace-by-trace comparison among these methods is provided in Fig. 4.17, which is randomly extracted from a 87.5% decimated seismic image. The result shows that our proposed approach delivers the highest reconstruction quality for the missing patterns.

4.4 Anti-Aliased Regular Seismic Data Interpolation by BTN

4.4.1 BTN with Spectrum Suppression for Aliasing Seismic Data Interpolation (BTN-SS)

Interpolating regularly sampled seismic data is usually more challenging than random scenar-

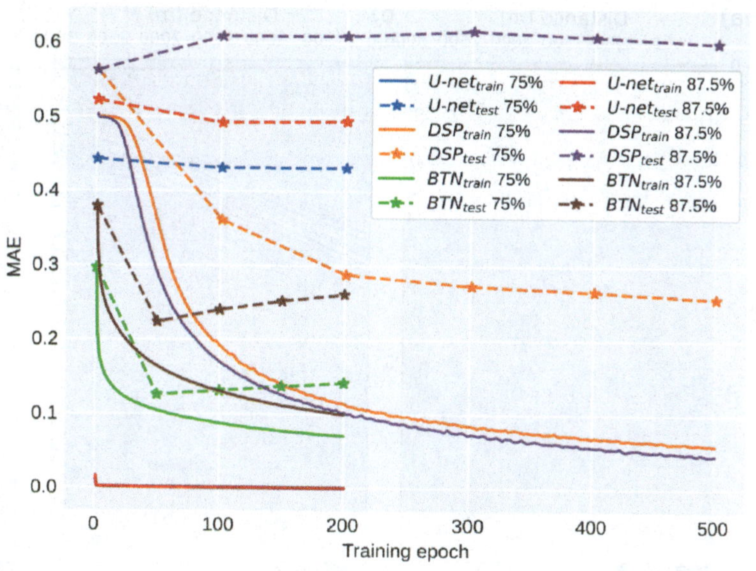

Fig. 4.11 Loss curve for randomly decimated Mobil AVO Viking Graben line 12 dataset. The decimation ratios are 75 and 87.5%, respectively

ios under the same decimation ratio. In random seismic data interpolation, critical information is more likely to be preserved by sufficient spatial sampling due to the randomness of the decimation patterns. However, this information could be missing in regular cases, causing unexpected problems and additional constraints are necessary to guarantee the desired results. While BTN successfully accomplishes the interpolation job for irregular decimated gathers, it fails to reconstruct regular decimated traces as illustrated in Fig. 4.18. An aliased result is produced by the BTN using the underlying self-supervised learning scheme, in which the network overfits on reconstructing only the known traces, ignoring the unknown testing traces. To address the ill-posedness issue, extra regularizations or constraints are necessary and we proposed an automatic f-k spectrum suppression method called BTN-SS.

Let $\mathcal{F}(\mathbf{x})$ denote the decimated common shot gathers in the f-k domain, where $\mathcal{F}$ symbolizes the 2D Fast Fourier Transform (2D-FFT). The unaliased seismic energy is assumed to be restricted in a particular area S in the f-k domain. Therefore, we can design an f-k mask $\mathbf{M} \in \mathcal{R}^{n_t \times n_r}$ by having $\mathbf{M}_i = 0$ if $i \in S$, and $\mathbf{M}_i = 1$ if $i \notin S$ to constrain the reconstruction of unaliased seismic components. The training loss can be updated as:

$$L_\theta(\mathbf{m}, \mathbf{M}) = \underbrace{\|\mathbf{x} - \sigma_\mathbf{m}(f_\theta(\mathbf{x}))\|}_{\text{misfit loss}} + \alpha \underbrace{\|\mathbf{M} \odot \mathcal{F}(f_\theta(\mathbf{x}))\|}_{\text{regularization loss}}, \quad (4.5)$$

where $\odot$ is the element-wise multiplication that attenuates the aliased energy outside the mask. The regularization term and original data misfit are weighted summed by α. A detector is used to estimate the shape of the spectrum mask, i.e. $S = \text{Detect}(\mathcal{F}(\mathbf{x}))$. In our work, we leverage a common assumption for seismic data in the f-k domain to shape the mask S as a cone-shaped area bounded by two dominant dips spanning from the zero point of the spectrum. Therefore, the objective is reduced to identifying the two boundaries automatically. We use the Hough transform method proposed by Duda and Hart (1972) to detect boundaries in our work, but it can be decided using any other techniques or even manually assigned in different scenarios. After adopting several

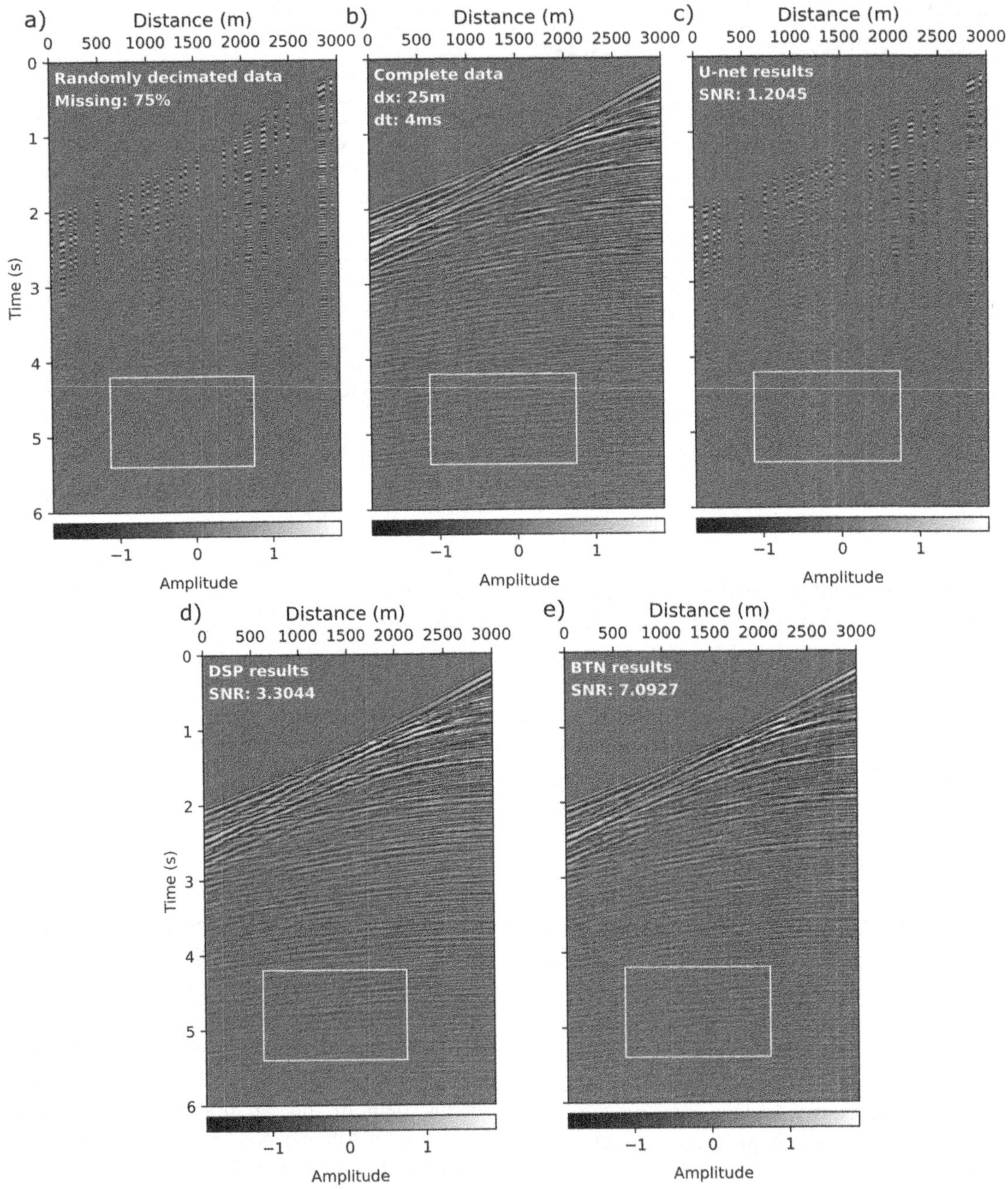

Fig. 4.12 Comparison for the randomly decimated field dataset with a missing ratio of 75%. (**a**) Decimated observation where the missing traces are inserted as 0s; (**b**) Complete seismic gathers; (**c**) Interpolated result by the naive U-Net method; (**d**) Interpolated result by the DSP method; (e) Interpolated result by the proposed BTN method

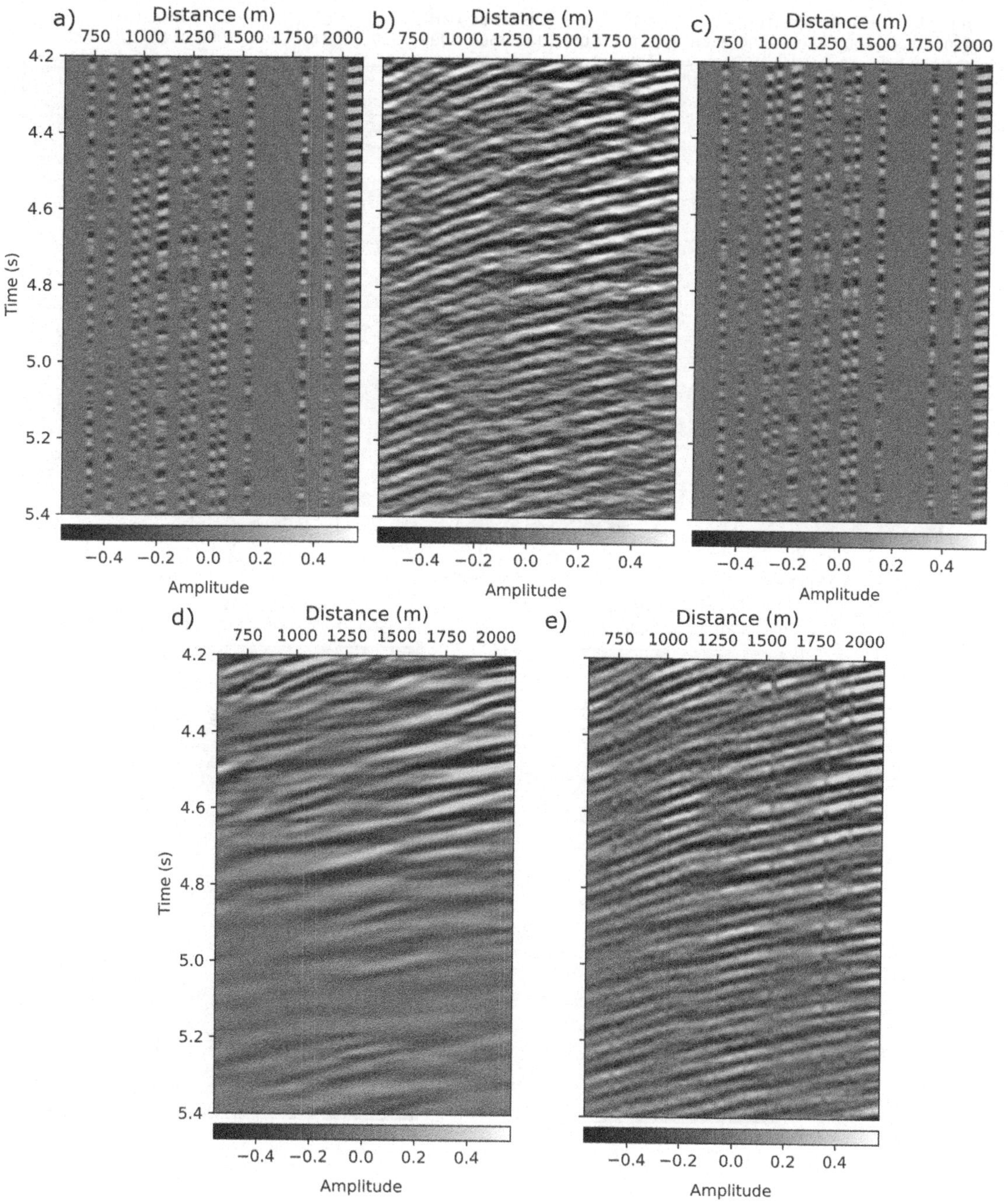

Fig. 4.13 Highlighted parts in the rectangular correspondingly in Fig. 4.12 (**a**)–(**e**)

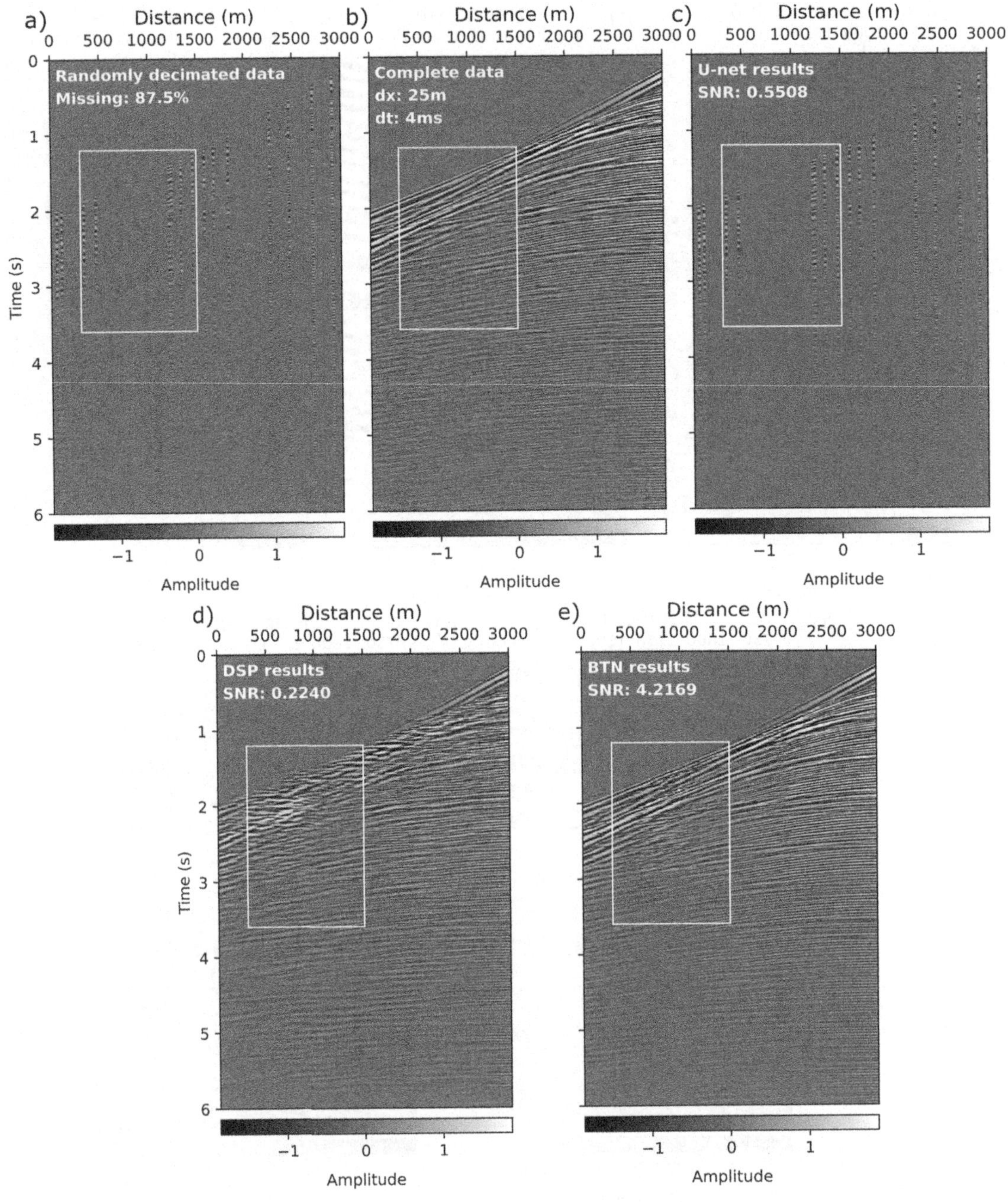

Fig. 4.14 Comparison for the randomly decimated field dataset with a missing ratio of 87.5%. (**a**) Decimated observation where the missing traces are inserted as 0s; (**b**) Complete seismic gathers; (**c**) Interpolated result by the naive U-Net method; (**d**) Interpolated result by the DSP method; (**e**) Interpolated result by the proposed BTN method

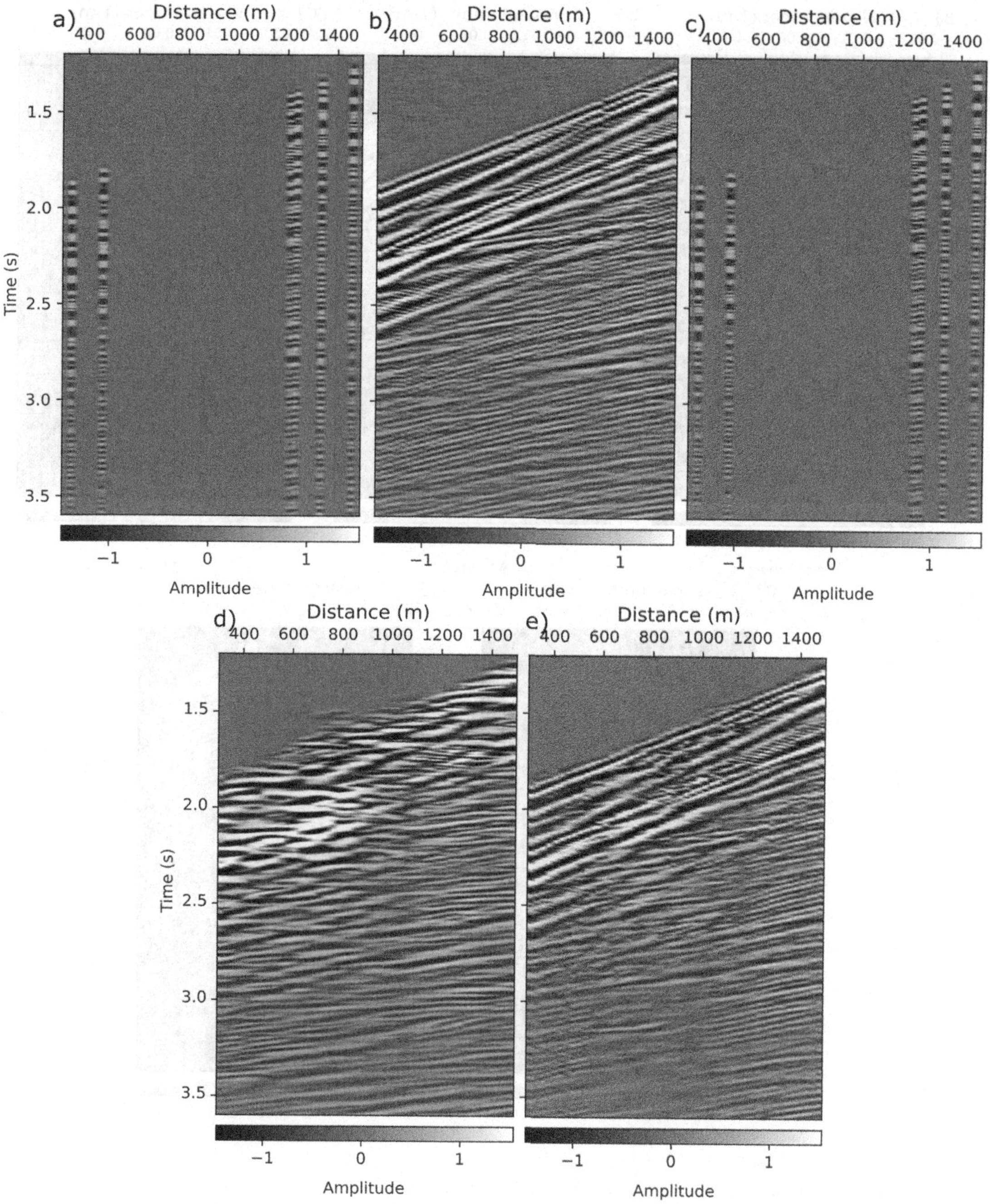

Fig. 4.15 Highlighted parts in the rectangular correspondingly in Fig. 4.14 (**a**)–(**e**)

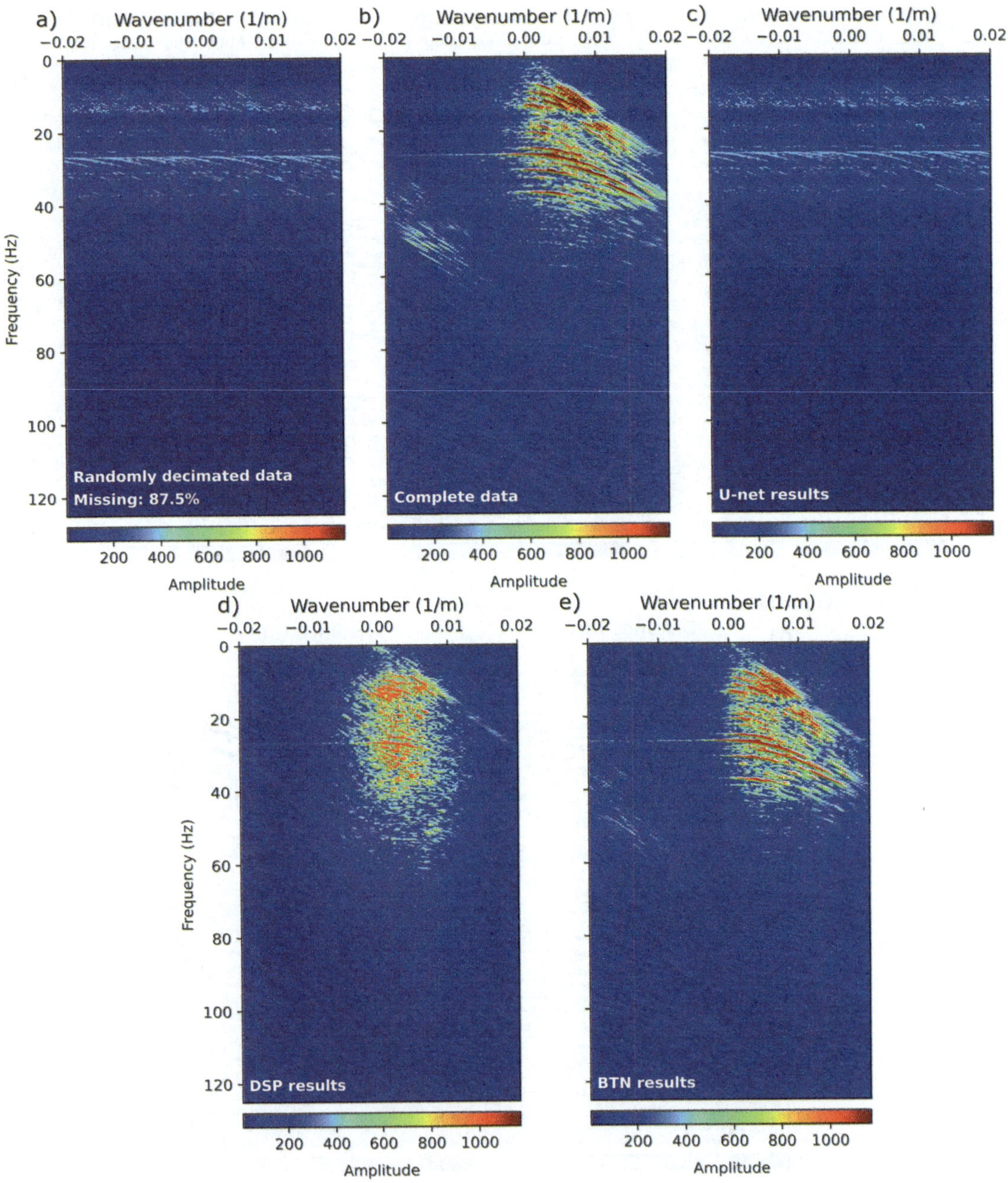

Fig. 4.16 F-k spectrum comparison for the randomly decimated field dataset with a missing ratio of 87.5%. (**a**) Decimated observation where the missing traces are inserted as 0s; (**b**) Complete seismic gathers; (**c**) Interpolated result by the naive U-Net method; (**d**) Interpolated result by the DSP method; (**e**) Interpolated result by the proposed BTN method

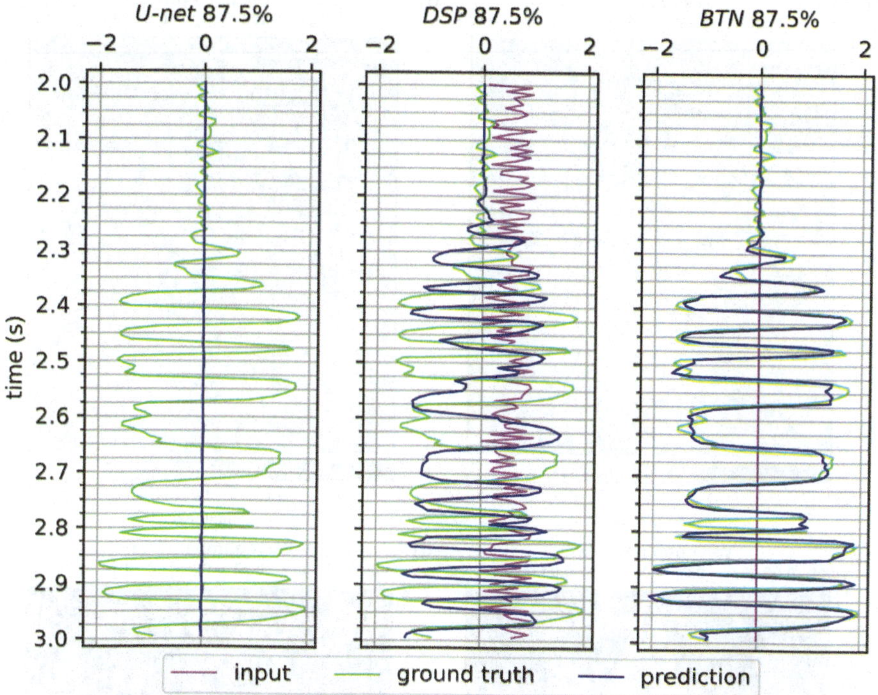

Fig. 4.17 Example traces of the field dataset with a random missing ratio of 87.5%

image processing steps, including amplitude balancing, dynamic thresholding with 98% percentile, erosion, and dilation, the two lines with the highest probability while crossing the zero point can be obtained. Such workflow using the SEAM–II unconventional data is illustrated in Fig. 4.19. This mask generation algorithm is fully automated and efficient for handling multiple seismic gathers. An appropriate hyper-parameter set for the boundary detection can be easily determined on one image and well-adaptive to the rest.

4.4.2 Mix-Training for Anti-Aliasing (BTN-Mix)

Another alternative method to the BTN-SS for tackling the aliasing issue in regular decimation scenarios is to incorporate an auxiliary decimated dataset and combine it with the original data to jointly train the BTN, called the BTN-Mix method. Although the proposed approach is a self-supervised learning method, overfitting can still occur when the missing patterns (indices of surrounding missing traces) for the undecimated data differ significantly from those of the decimated traces. This is because the training process for BTN is fully based on undecimated traces, and the inference stage is performed on missing traces that have never been supervised. This train-test variation is problematic especially for regularly decimated traces. To address this issue, introducing randomness to diversify the missing patterns becomes essential. Instead of creating an entirely new dataset with random decimation patterns, our strategy involves reusing the target dataset and sorting it into the common receiver domain (assuming the original interpolation occurs in the common shot domain). By randomly decimating

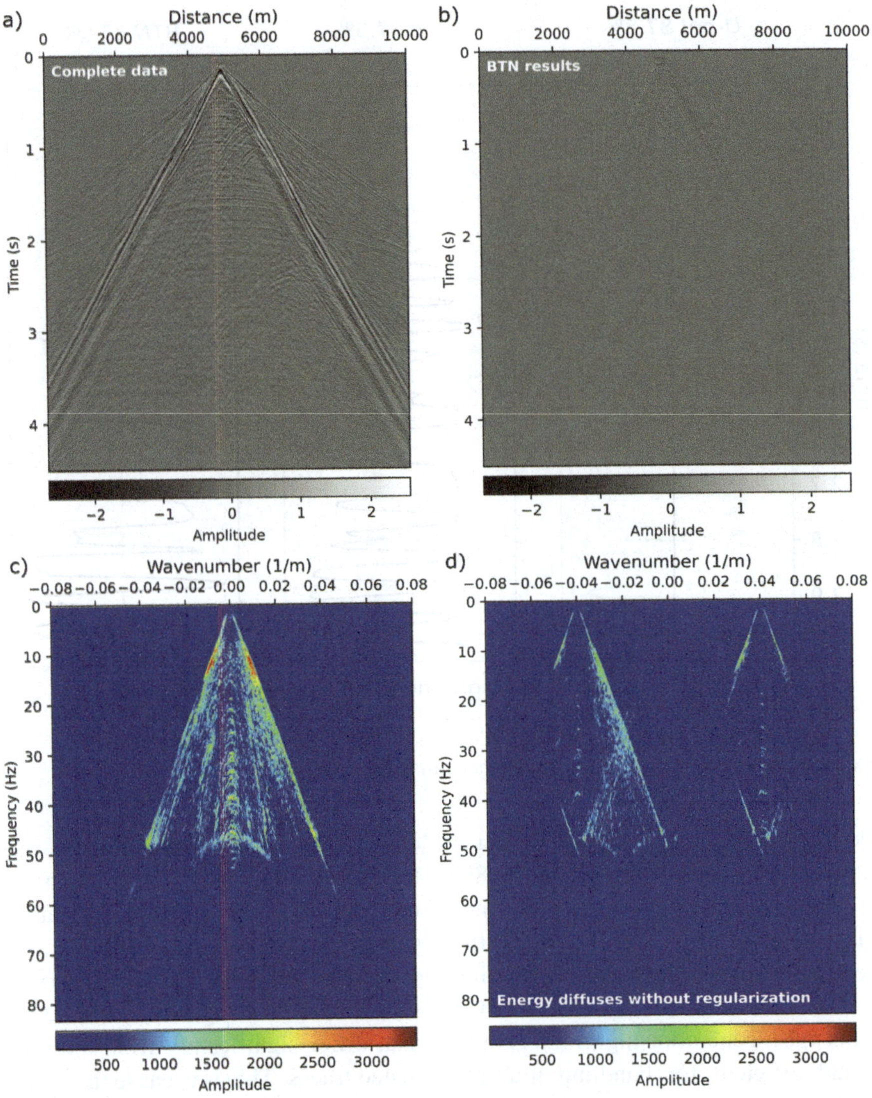

Fig. 4.18 BTN fails by giving a trivial answer for regularly decimated data with a 75% decimation ratio

the complete common receiver gathers with the same decimation ratio and incorporating them into the training dataset, we ensure a high degree of similarity between the target and auxiliary datasets. This similarity is underpinned by the reciprocal theory, guaranteeing maximum resemblance. The network then undergoes training using these mixed mini-batches, with the training loss formulated as per Eq. (4.2).

4.4.3 Experimental Results for Regular Seismic Data Interpolation

We evaluate the proposed anti-aliasing regular interpolation method on the SEAM–II unconventional (synthetic) and Mobil AVO Viking Graben line 12 (field) datasets. The quantitative comparison for the SEAM–II unconventional data is pre-

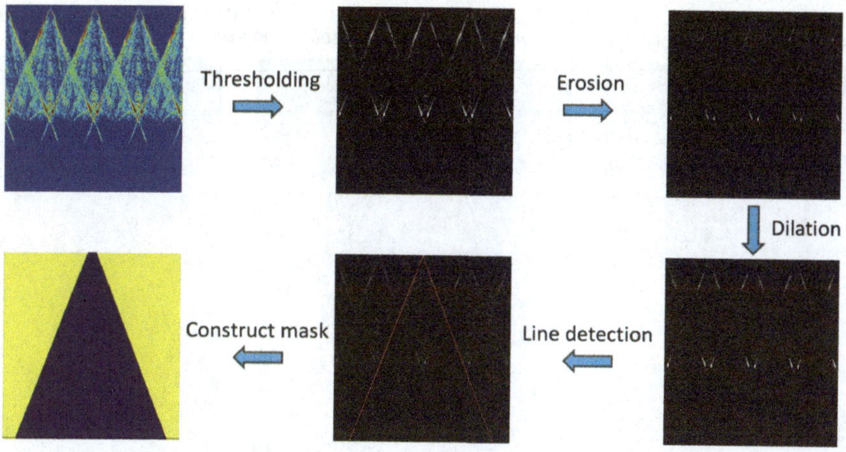

Fig. 4.19 The workflow of generating the f-k mask for the spectrum suppression constraint by pre-processing and the Hough transform line detection method

Table 4.3 Metrics for regularly interpolating SEAM–II unconventional dataset

Methods	Missing ratio	MAE	MSE	SNR	Epochs
DSP	50%	0.1227	0.0377	7.0195	800
BTN-SS	50%	**0.0312**	**0.0165**	**10.6159**	200
BTN-Mix	50%	0.0790	0.0329	7.6110	200
DSP	75%	0.1396	0.0553	5.3353	800
BTN-SS	75%	0.0994	0.0874	3.3422	200
BTN-Mix	75%	**0.0790**	**0.0329**	**7.6110**	200

The bold values highlights the best results

sented in Table 4.3. BTN-SS delivers the best performance under the 50% missing ratio, whereas BTN-Mix outperforms all other approaches in the 75% decimation scenario. Furthermore, the proposed BTN-SS and BTN-Mix approaches exhibit faster convergence speeds compared to the baseline DSP method, with an epoch ratio of 1:4.

The interpolated results for SEAM-II unconventional dataset under a 50% regular decimation ratio are presented in Figs. 4.20 and 4.21. The network is trained with a total of 200 common shot gathers for all approaches. The first row shows the whole seismic gathers and the second row illustrates the zoom-in plots in the emphasized regions by the rectangles. It is evident that the proposed BTN-SS and BTN-Mix solutions remarkably outperform the DSP approach both qualitatively and quantitatively. As demonstrated in the enlarged visualization in the second row, the proposed two methods have higher accuracy for the amplitudes and most details in the bottom region are successfully recovered. Only a minor difference is observed between the BTN-SS and BTN-Mix results. However, the DSP method fails to reconstruct the weak events in late arrivals.

Figures 4.22, 4.23, and 4.24 demonstrates the result comparison when the decimation ratio increases to 75% on the same synthetic dataset, and Fig. 4.25 depicts the corresponding f-k spectra. In this case, the BTN-SS and BTN-Mix methods consistently outperform the DSP method in reconstructing complex events for the late arrivals in the third row of Fig. 4.24. However, the BTN-SS method is challenged by the aliasing problem at the steepest dips in the second row of Fig. 4.23. Due to the increased missing ratio, critical information is missing in the near-offset regions which cannot be learned by the BTN from the under-sampled input. However, the BTN-Mix method successfully prevents the aliasing issue in Fig. 4.23e, yielding precise predictions for both the top regions and bottom regions. This phenomenon illustrates the effectiveness of incorporating appropriate auxiliary datasets for training

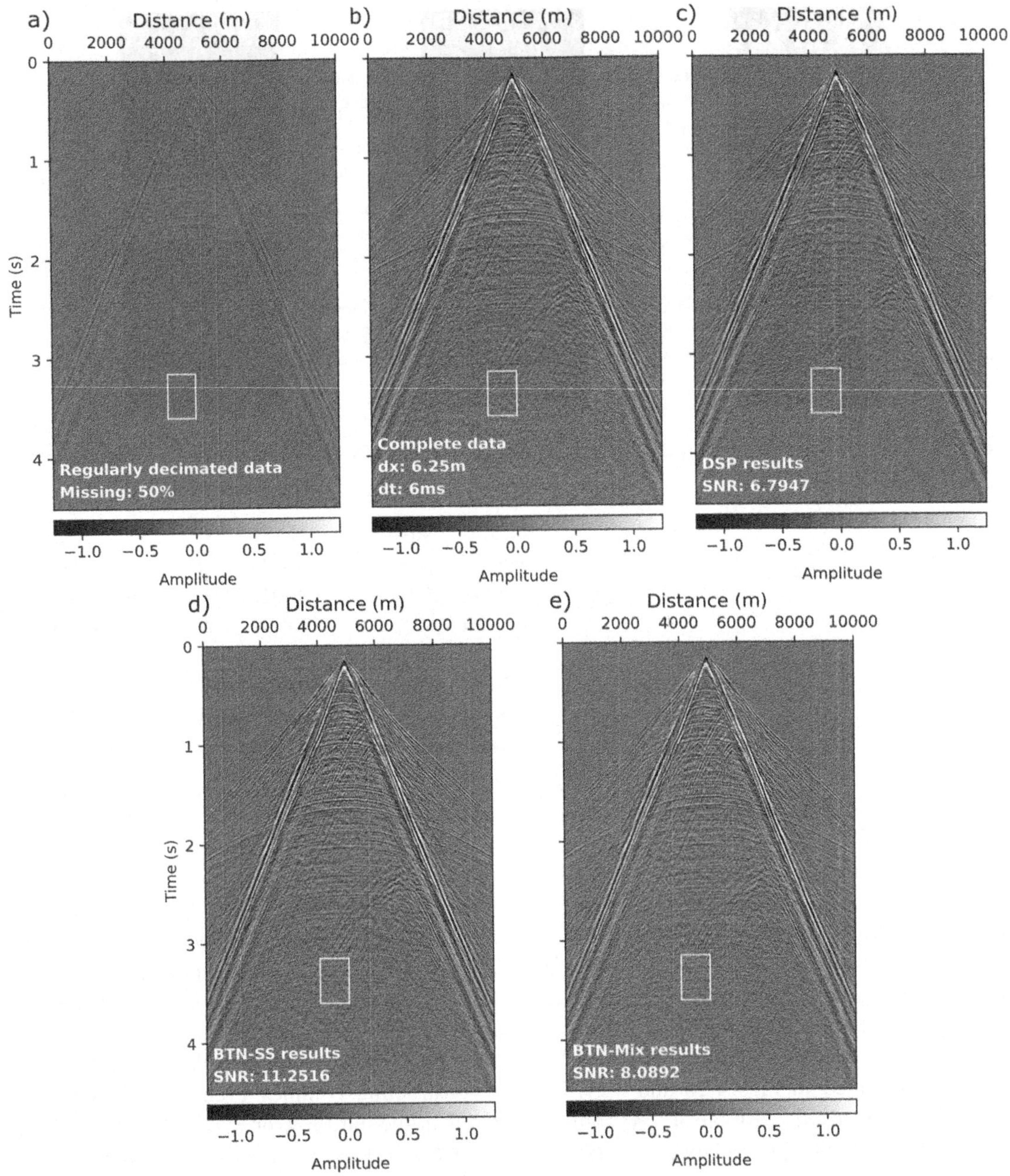

Fig. 4.20 Comparison for the regularly decimated SEAM–II unconventional dataset with a missing ratio of 50%. (**a**) Decimated observation where the missing traces are inserted as 0s; (**b**) Complete seismic gathers; (**c**) Interpolated result by the DSP method; (**d**) Interpolated result by the BTN-SS method; (**e**) Interpolated result by the BTN-Mix method

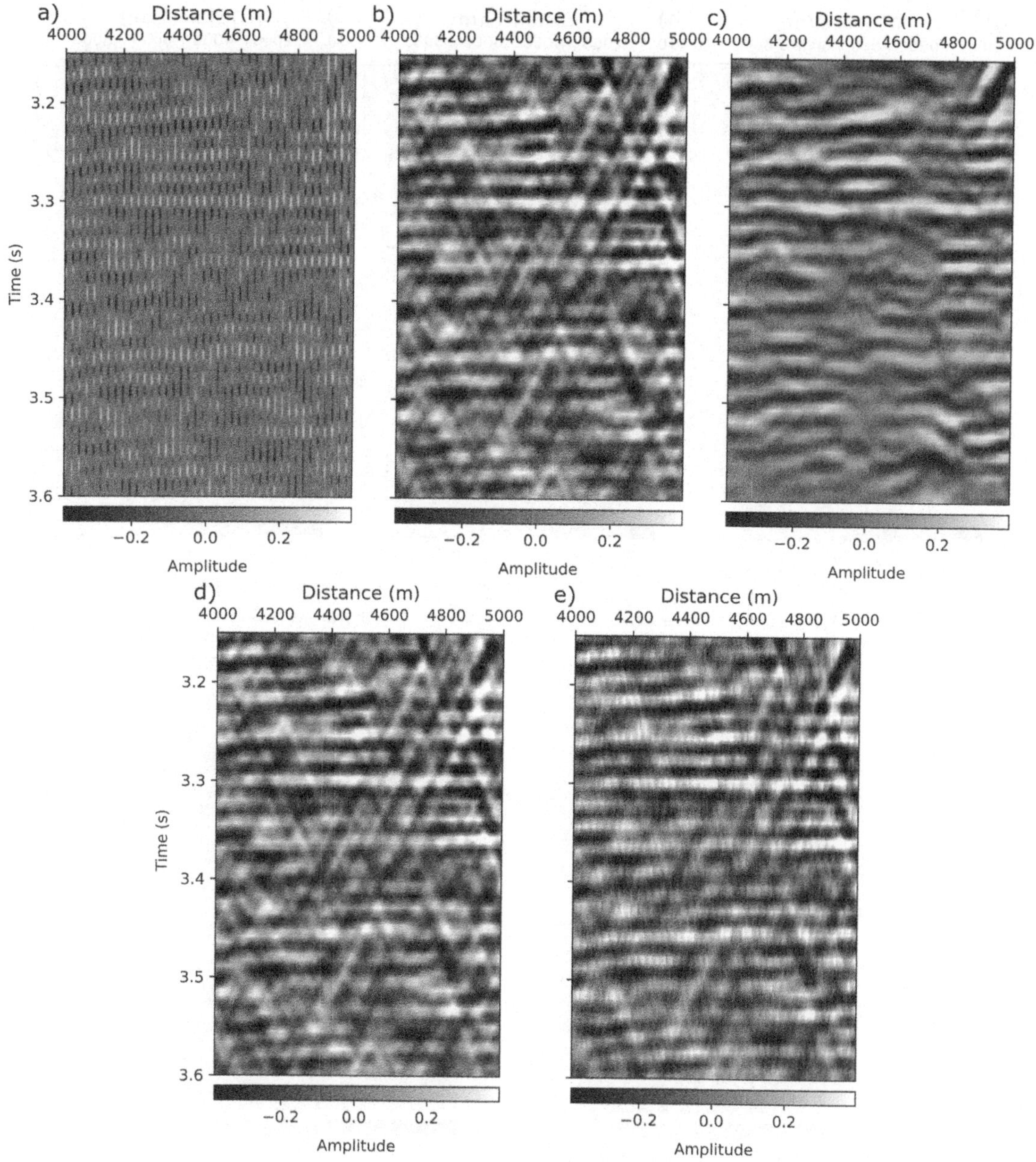

Fig. 4.21 Highlighted section in the rectangular correspondingly in Fig. 4.20 (**a**)–(**e**)

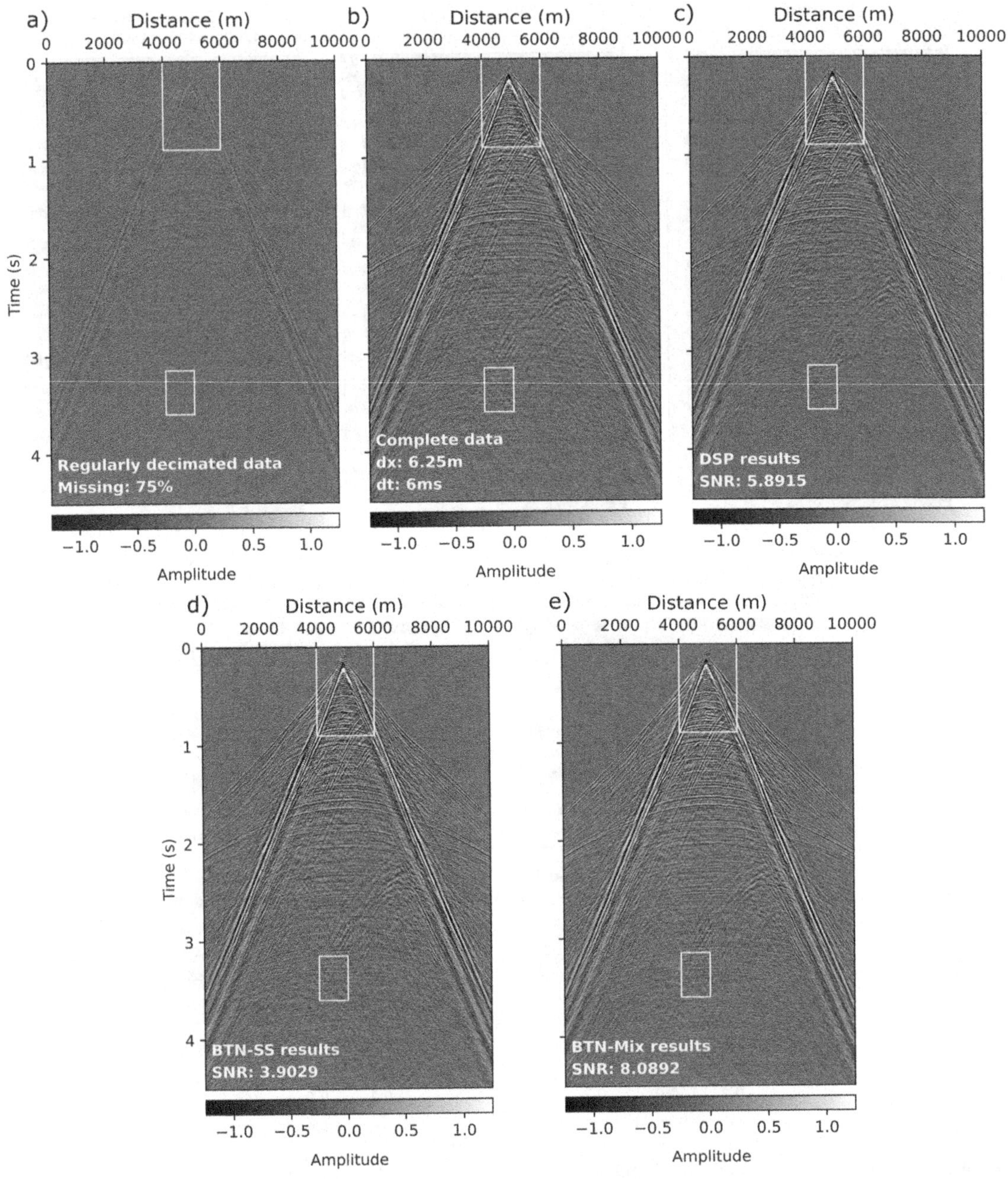

Fig. 4.22 Comparison for the regularly decimated SEAM–II unconventional dataset with a missing ratio of 75%. (**a**) Decimated observation where the missing traces are inserted as 0s; (**b**) Complete seismic gathers; (**c**) Interpolated result by the DSP method; (**d**) Interpolated result by the BTN-SS method; (**e**) Interpolated result by the BTN-Mix method

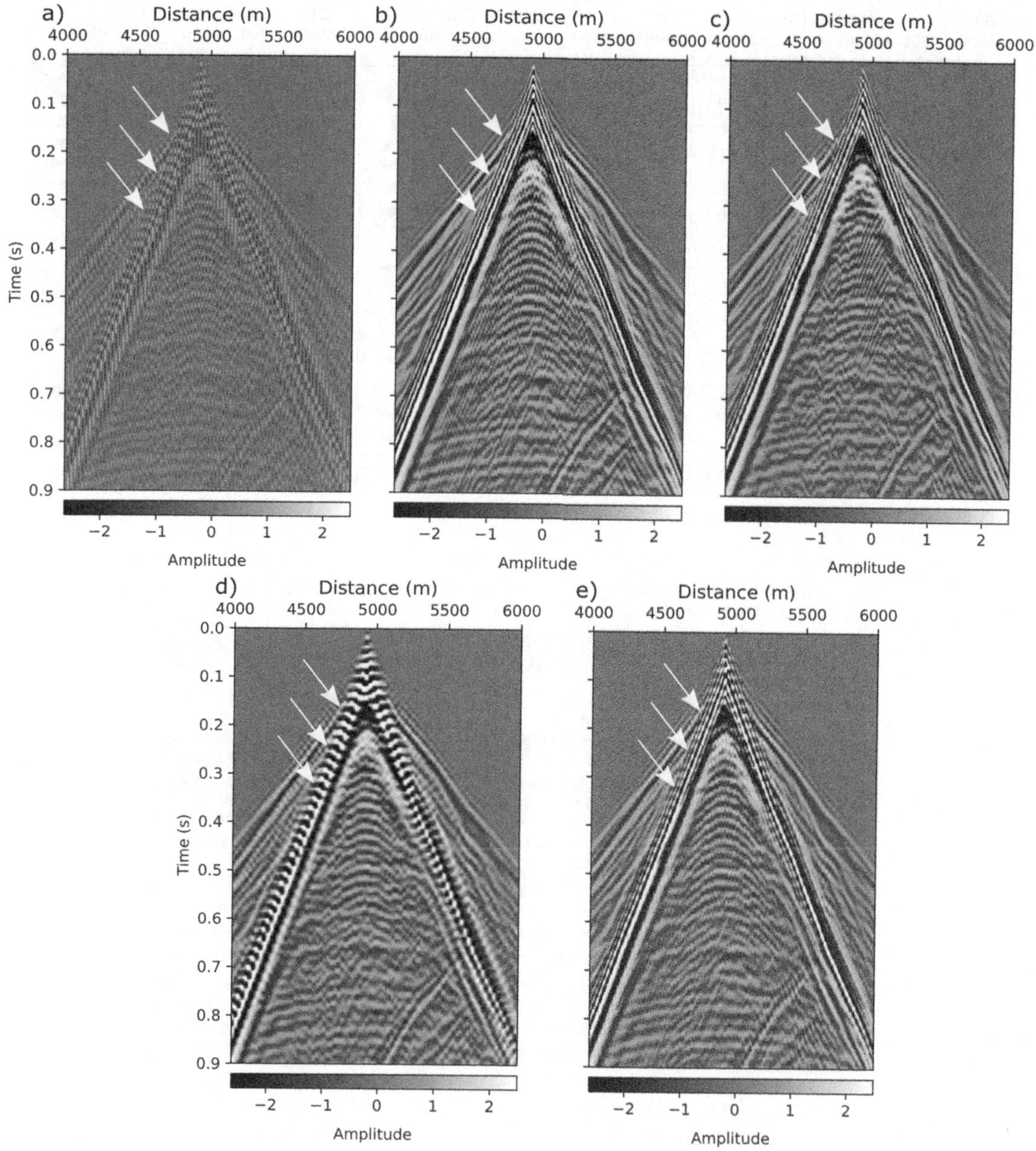

Fig. 4.23 Highlighted section in the shallow rectangular correspondingly in Fig. 4.22 (**a**)–(**e**)

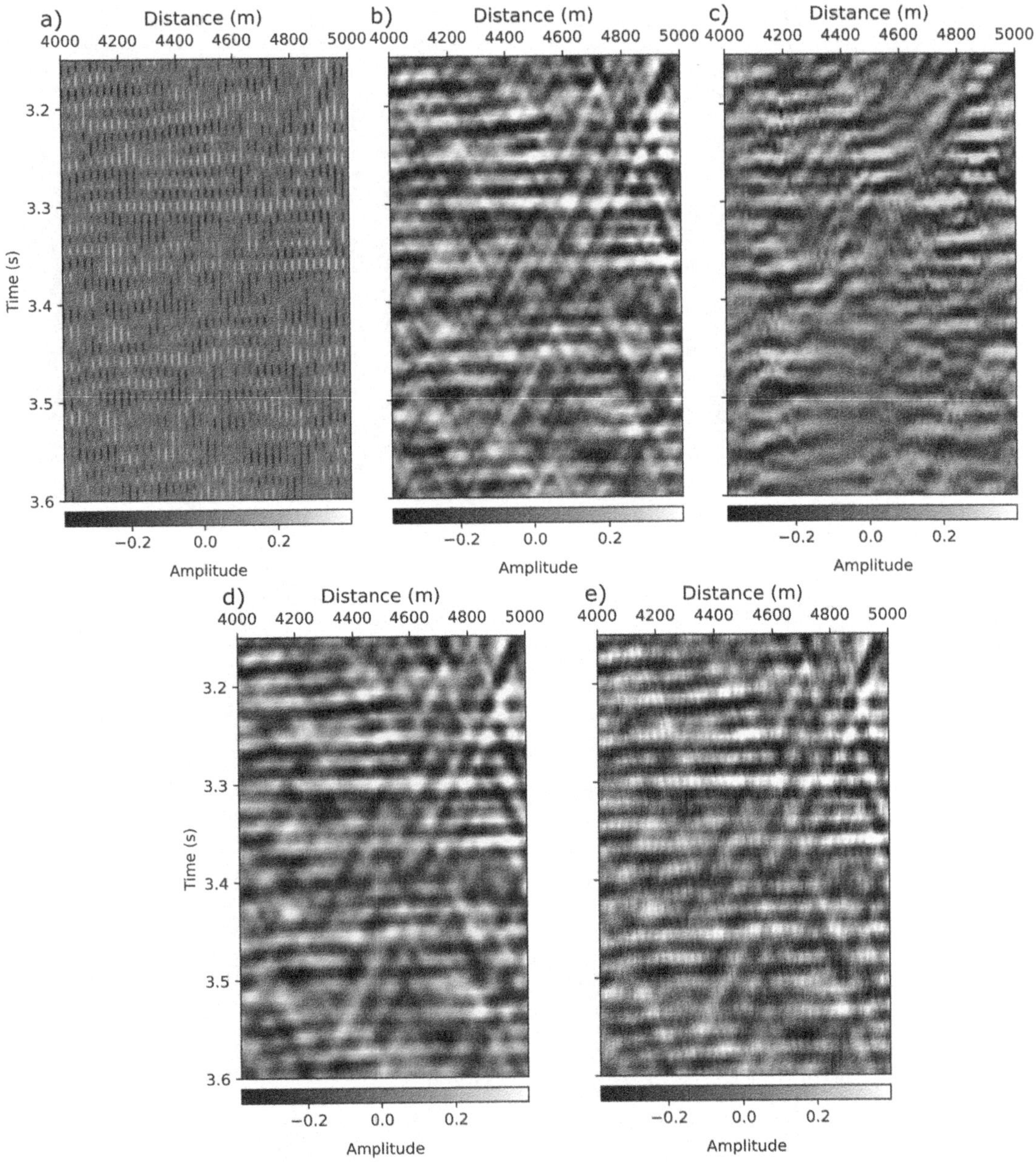

Fig. 4.24 Highlighted section in the deep rectangular correspondingly in Fig. 4.22 (**a**)–(**e**)

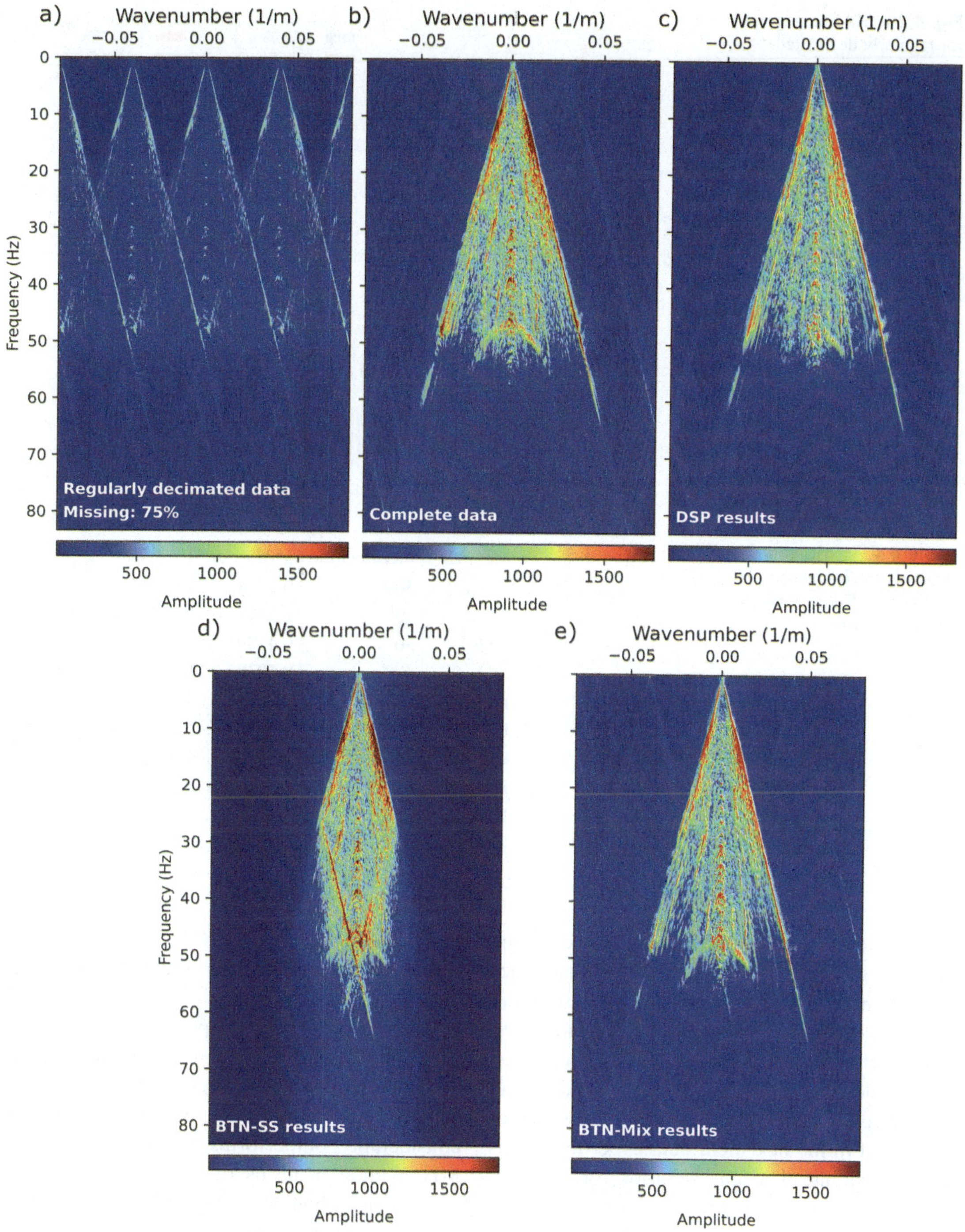

Fig. 4.25 F-k spectrum comparison for the regularly decimated SEAM–II unconventional dataset with a missing ratio of 75%. (**a**) Decimated observation where the missing traces are inserted as 0s; (**b**) Complete seismic gathers; (**c**) Interpolated result by the DSP method; (**d**) Interpolated result by the BTN-SS method; (**e**) Interpolated result by the BTN-Mix method

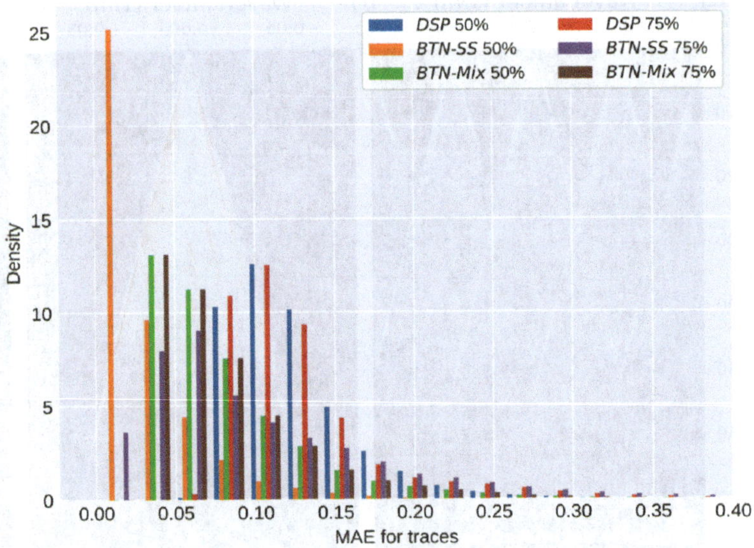

Fig. 4.26 MAE histogram for regularly decimated SEAM–II unconventional data. Under a 50% decimation ratio

Table 4.4 Metrics for regularly interpolating Mobil AVO Viking Graben line 12 dataset

Methods	Missing ratio	MAE	MSE	SNR	Epochs
U-net	50%	0.2518	0.2499	3.0046	<100
DSP	50%	0.3334	0.3103	2.1062	500
BTN	50%	0.4109	0.5446	−0.3692	200
BTN-SS	50%	**0.1939**	**0.1111**	**6.5626**	200
BTN-Mix	50%	0.2294	0.2357	3.2765	200
U-net	75%	0.3743	0.3744	1.2465	<100
DSP	75%	0.4015	0.5421	−0.3669	500
BTN	75%	0.4309	0.4626	0.3219	200
BTN-SS	75%	**0.2768**	**0.2346**	**3.2955**	200
BTN-Mix	75%	0.3981	0.5363	-0.3010	200

The bold values highlights the best results

the network. On the other hand, the common receiver gathers can be the best candidate since it is often available and share the same structure and amplitude distributions as the target dataset. The DSP method does not have the strong aliasing problem in Fig. 4.23c due to the randomness introduced in the input noise image. However, as presented in the next field data experiment, this is not guaranteed in all scenarios for the DSP method when the data structure varies.

The histogram for the MAE comparison is illustrated in Fig. 4.26 under both 50 and 75% decimation ratios for the SEAM-II unconventional data. Typically, the increase in the decimation ratio from 50 to 75% will result in a larger error for the network. Overall, the BTN-SS method achieves the best performance for 50% decimation scenario, while the BTN-Mix approach outperform the others for the 65% decimated dataset.

We then evaluate the proposed methods on the field dataset (Mobile AVO Viking Graben line 12), with numerical comparisons depicted in Table 4.4. In this case, the BTN-SS method considerably outperforms the DSP and BTN-Mix method in terms of both MAE and SNR metrics under both 50 and 75% decimation ratios. Although the BTN-Mix approach provides the second-best result, a significant gap remains between its performance and that of the top-performing BTN-SS approach. The U-net, DSP, and BTN-Mix approaches are roughly at the same level, but the BTN-alone fails for all regular cases.

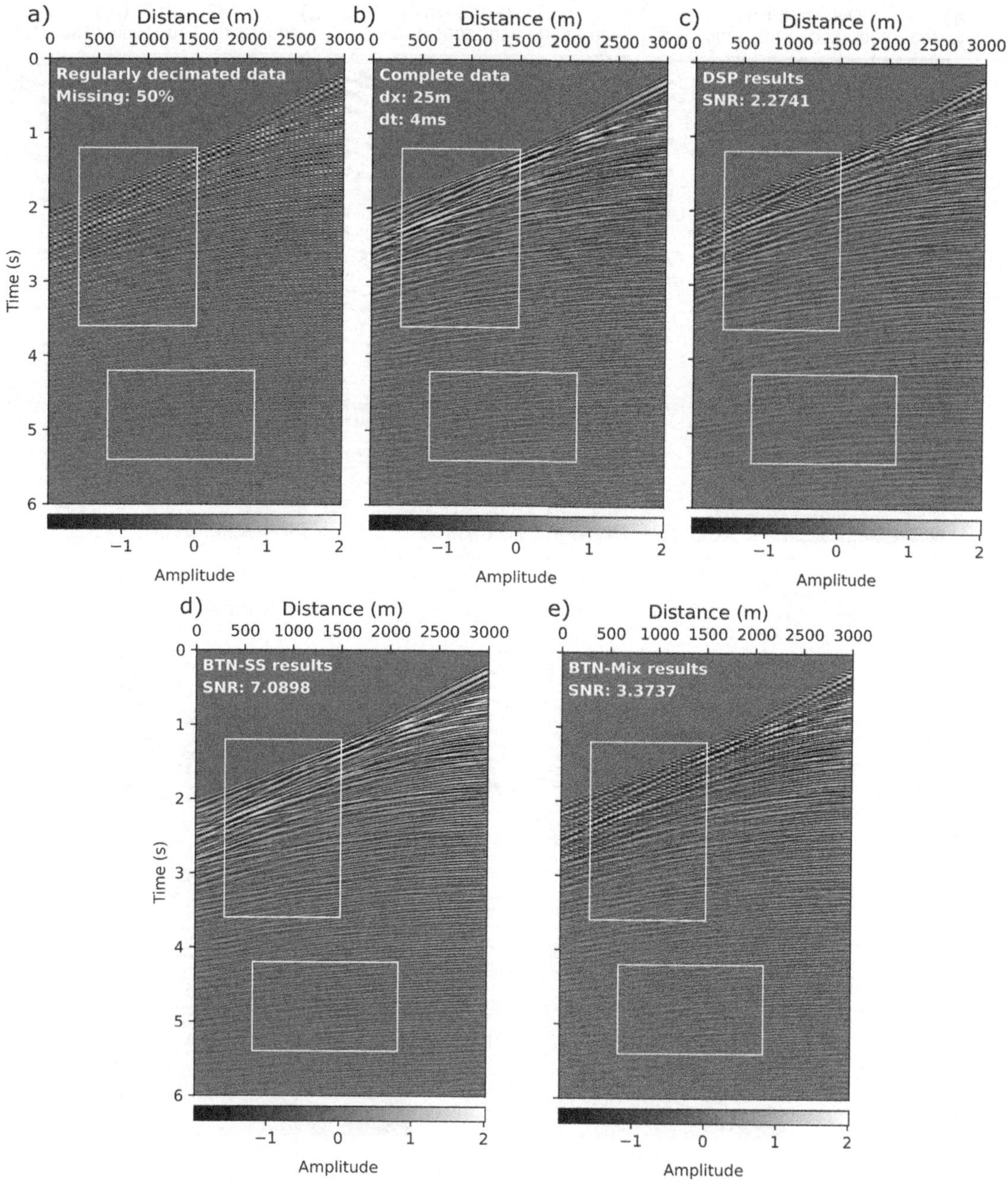

Fig. 4.27 Comparison for the regularly decimated Mobile AVO Viking Graben line 12 dataset with a missing ratio of 50%. (**a**) Decimated observation where the missing traces are inserted as 0s; (**b**) Complete seismic gathers; (**c**) Interpolated result by the DSP method; (**d**) Interpolated result by the BTN-SS method; (**e**) Interpolated result by the BTN-Mix method

The interpolated results for the DSP, BTN-SS, and BTN-Mix approaches with 50 and 75% of the total gathers decimated are presented in Figs. 4.27, 4.28, 4.29, 4.30, 4.31, and 4.32, respectively. Compared to the SEAM synthetic dataset, the field data exhibits steeper dips, resulting in more pronounced aliasing issues. Figure 4.27c, e demonstrate that the DSP and BTN-

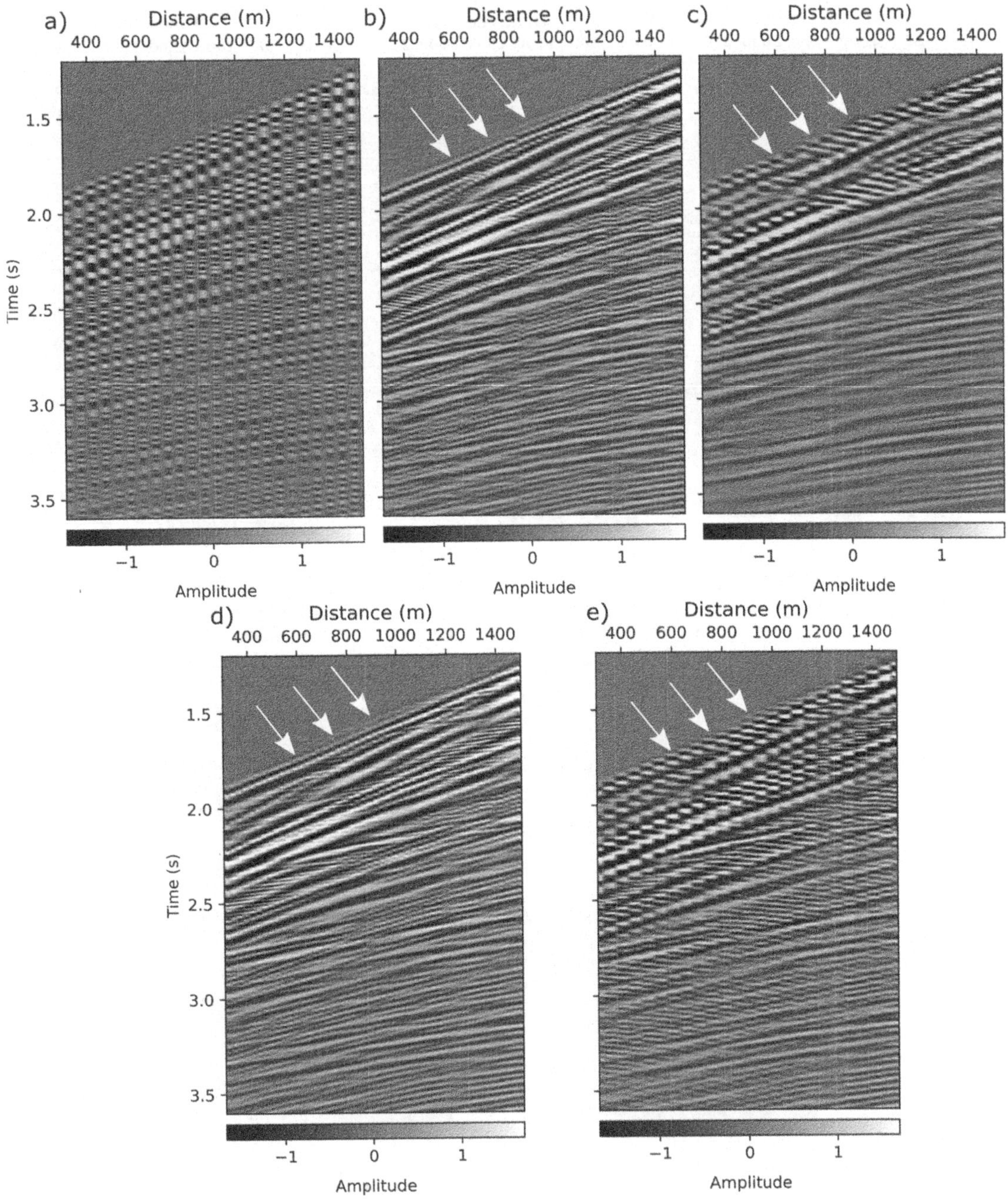

Fig. 4.28 Highlighted section in the shallow rectangular correspondingly in Fig. 4.27 (**a**)–(**e**)

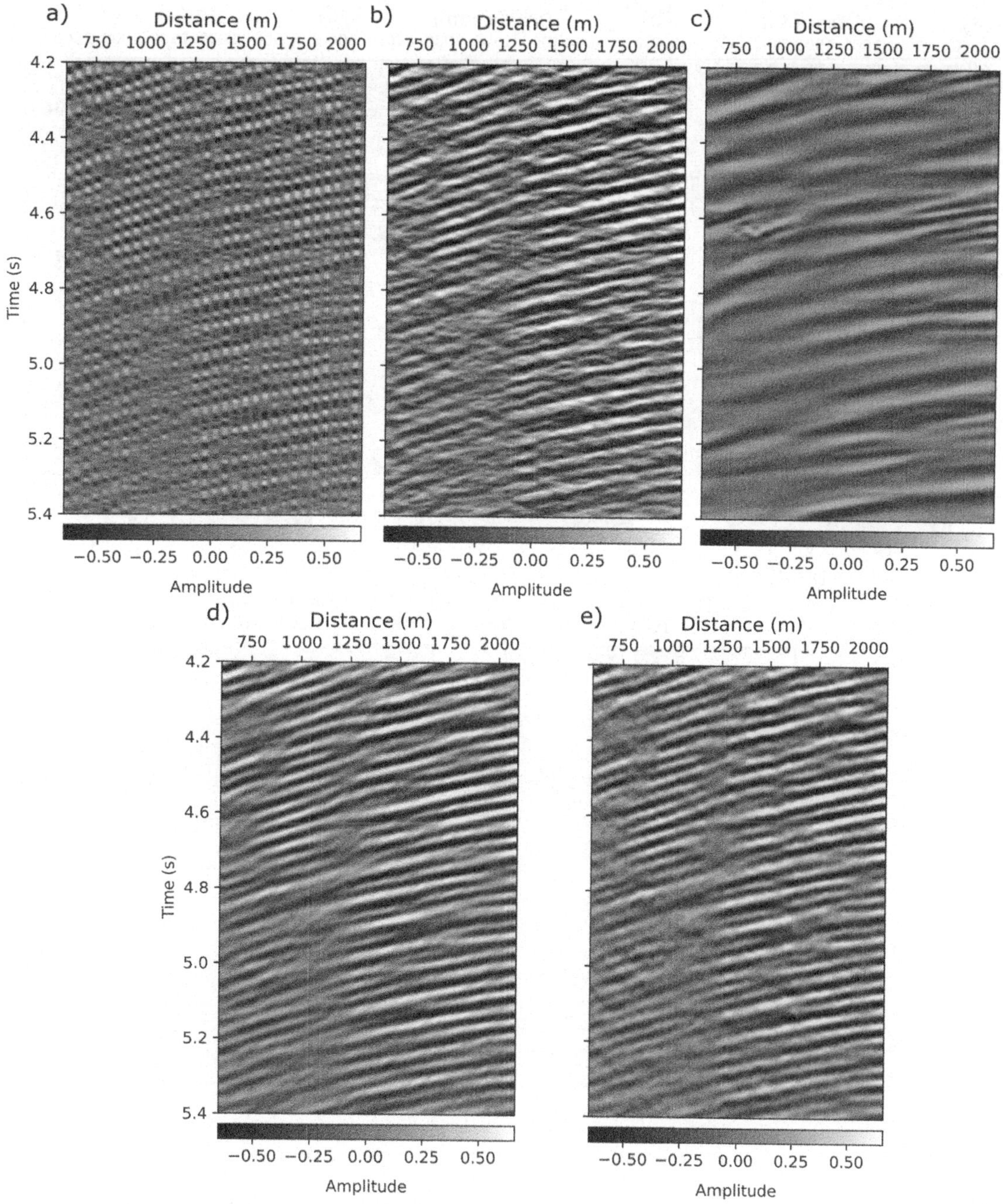

Fig. 4.29 Highlighted section in the deep rectangular correspondingly in Fig. 4.27 (**a**)–(**e**)

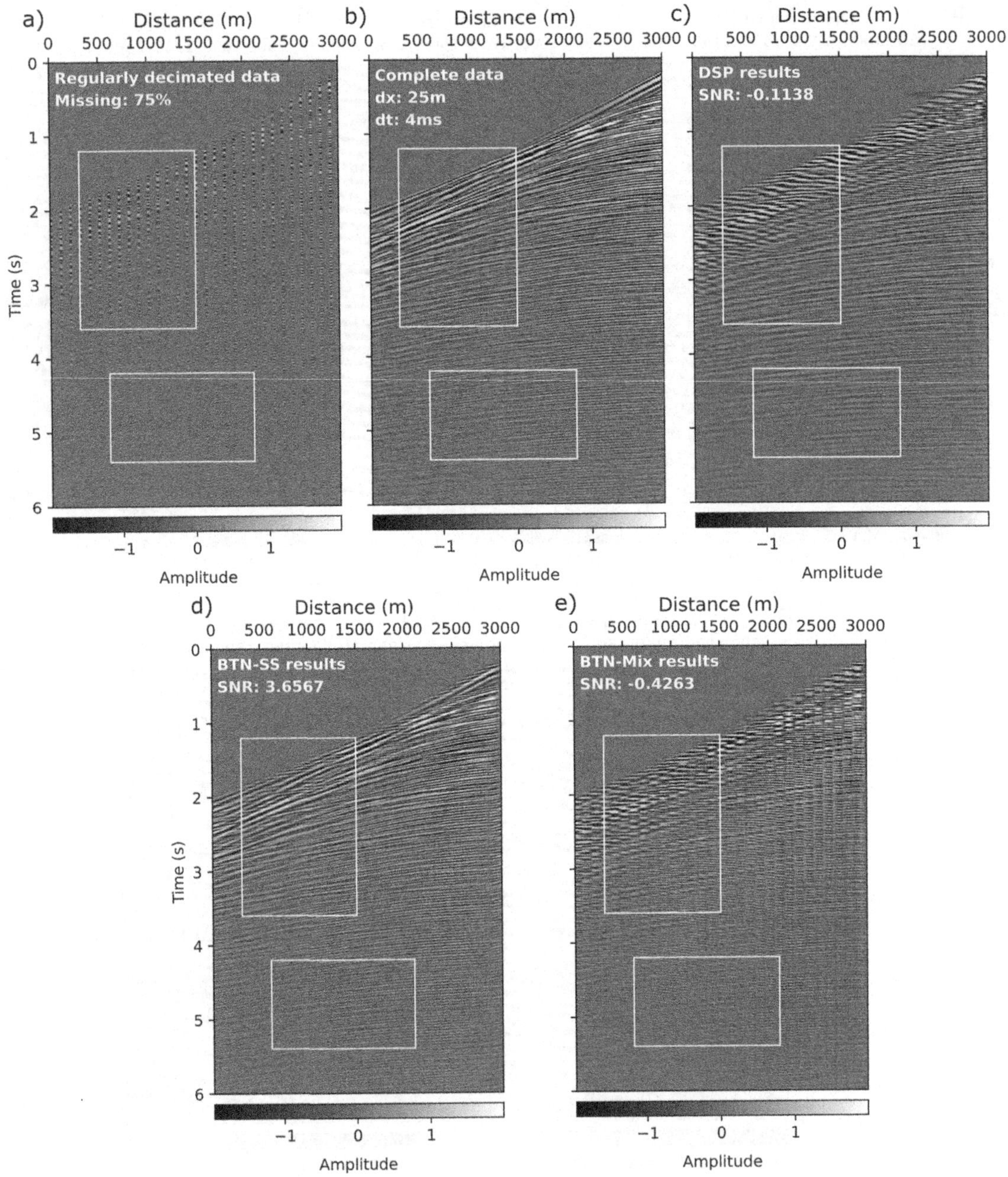

Fig. 4.30 Comparison for the regularly decimated Mobile AVO Viking Graben line 12 dataset with a missing ratio of 75%. (**a**) Decimated observation where the missing traces are inserted as 0s; (**b**) Complete seismic gathers; (**c**) Interpolated result by the DSP method; (**d**) Interpolated result by the BTN-SS method; (**e**) Interpolated result by the BTN-Mix method

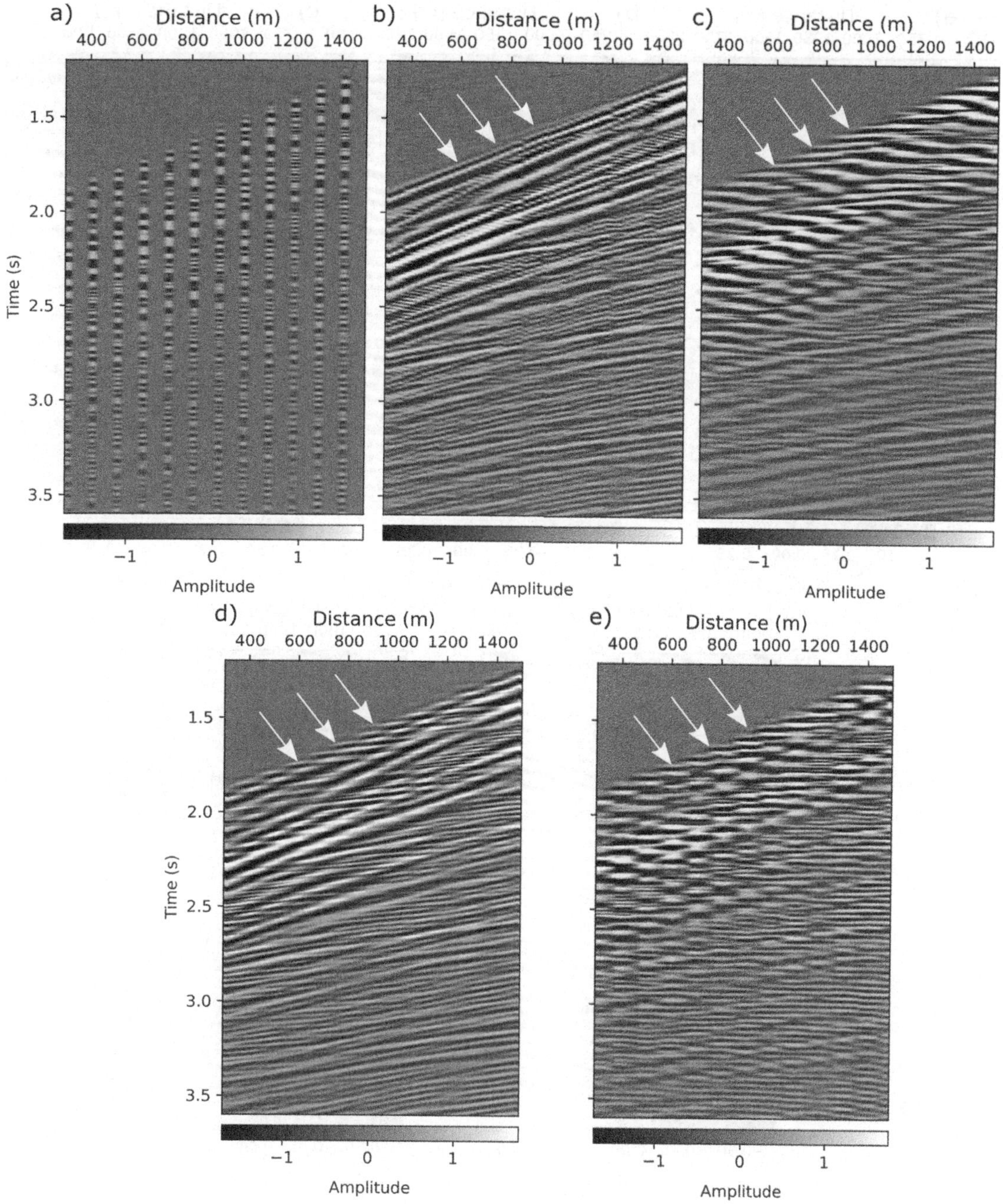

Fig. 4.31 Highlighted section in the shallow rectangular correspondingly in Fig. 4.30 (**a**)–(**e**)

Mix methods produce reasonable results, reconstructing low-frequency and low-wavenumber components to a certain degree. Nevertheless, only the BTN-SS approach can successfully generate unaliased results, recovering and reconstructing high-frequency and high-wavenumber signals. This observation further highlights the performance gap between the BTN-SS method and others in Table 4.4. In the 75% decimation scenario presented in Figs. 4.30, 4.31, and 4.32

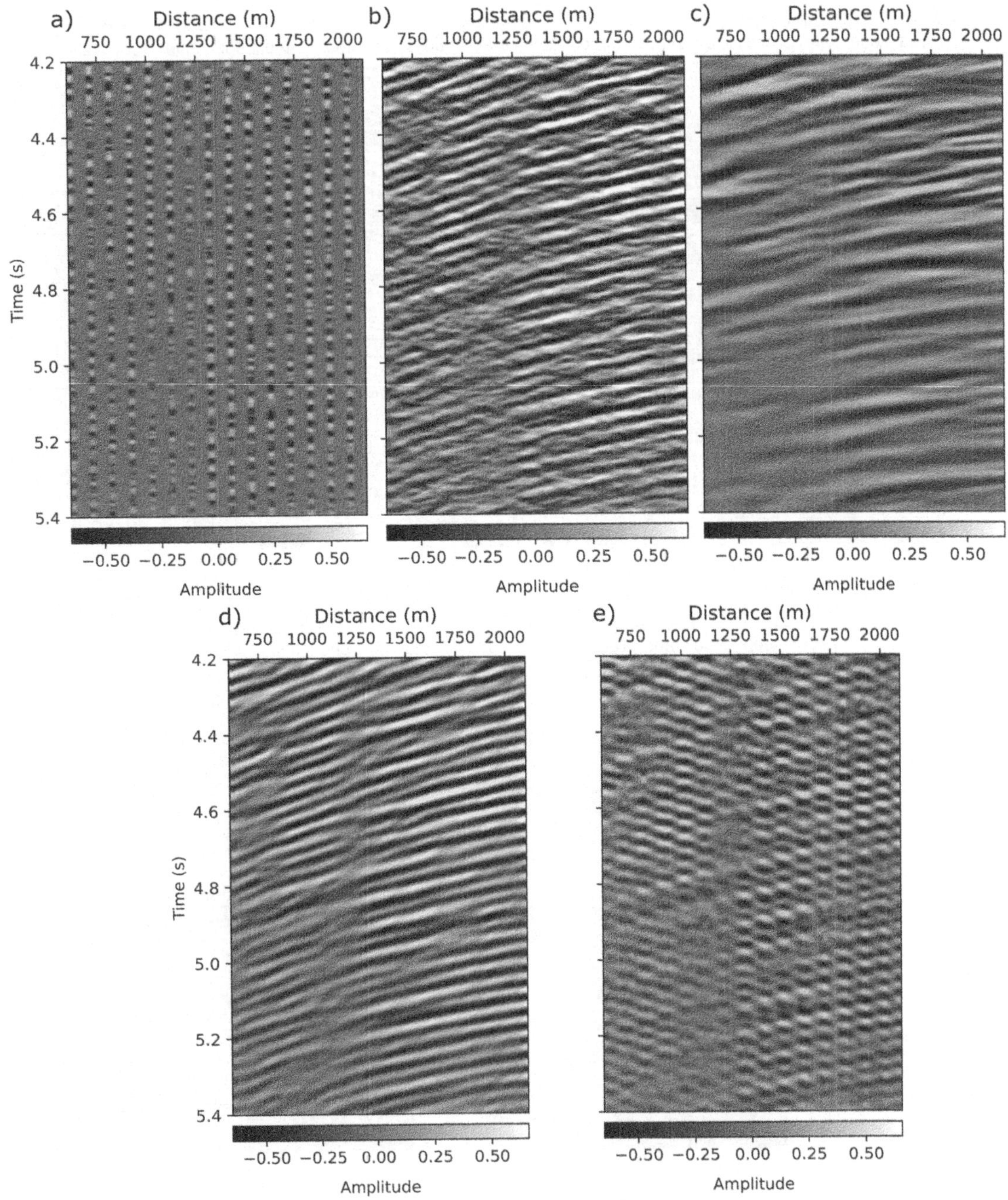

Fig. 4.32 Highlighted section in the deep rectangular correspondingly in Fig. 4.30 (**a**)–(**e**)

the aliasing problem intensifies. The BTN-SS method effectively addresses the aliasing issue and accurately reconstructs most events in both early and late arrivals. In contrast, the DSP and BTN-Mix methods produce exacerbated aliasing problems. The Fourier domain visualizations in Figs. 4.33 and 4.34 offer additional insight into these comparisons. Under the 50% decimation case, the f-k spectrum of the decimated gather (Fig. 4.33a) is not severely aliased, and all three

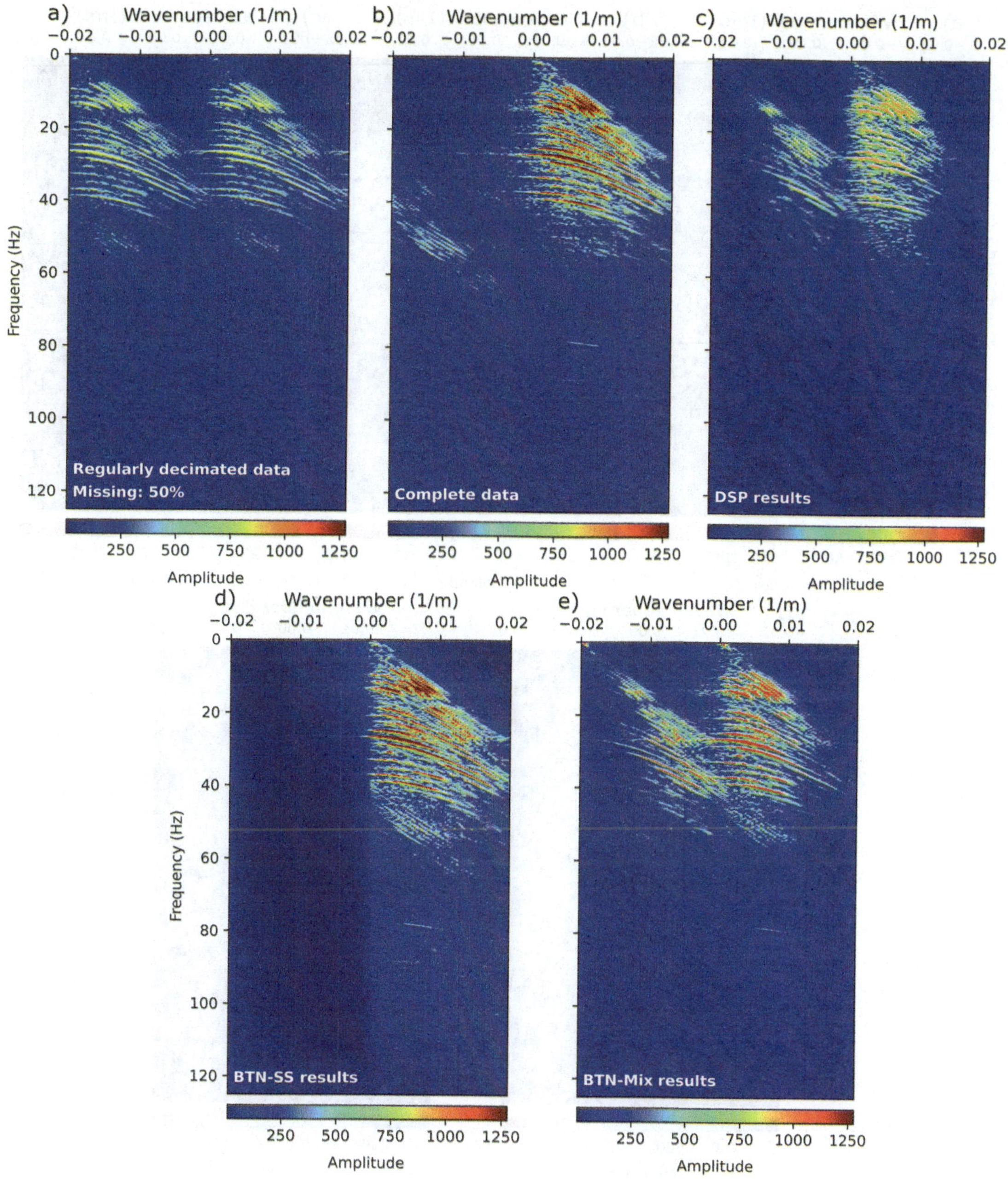

Fig. 4.33 F-k spectrum comparison for the regularly decimated Mobile AVO Viking Graben line 12 dataset with a missing ratio of 50%. (**a**) Decimated observation where the missing traces are inserted as 0s; (**b**) Complete seismic gathers; (**c**) Interpolated result by the DSP method; (**d**) Interpolated result by the BTN-SS method; (**e**) Interpolated result by the BTN-Mix method

methods can restore the majority of the unaliased area, but with a slight aliased portion in the DSP and BTN-Mix results (Fig. 4.33c, e). As the aliasing problem deteriorates in Fig. 4.34a, the DSP and BTN-Mix methods completely fail, only restoring aliased energy. This indicates that the networks tend to reconstruct low-frequency and low-wavenumber components instead of complex

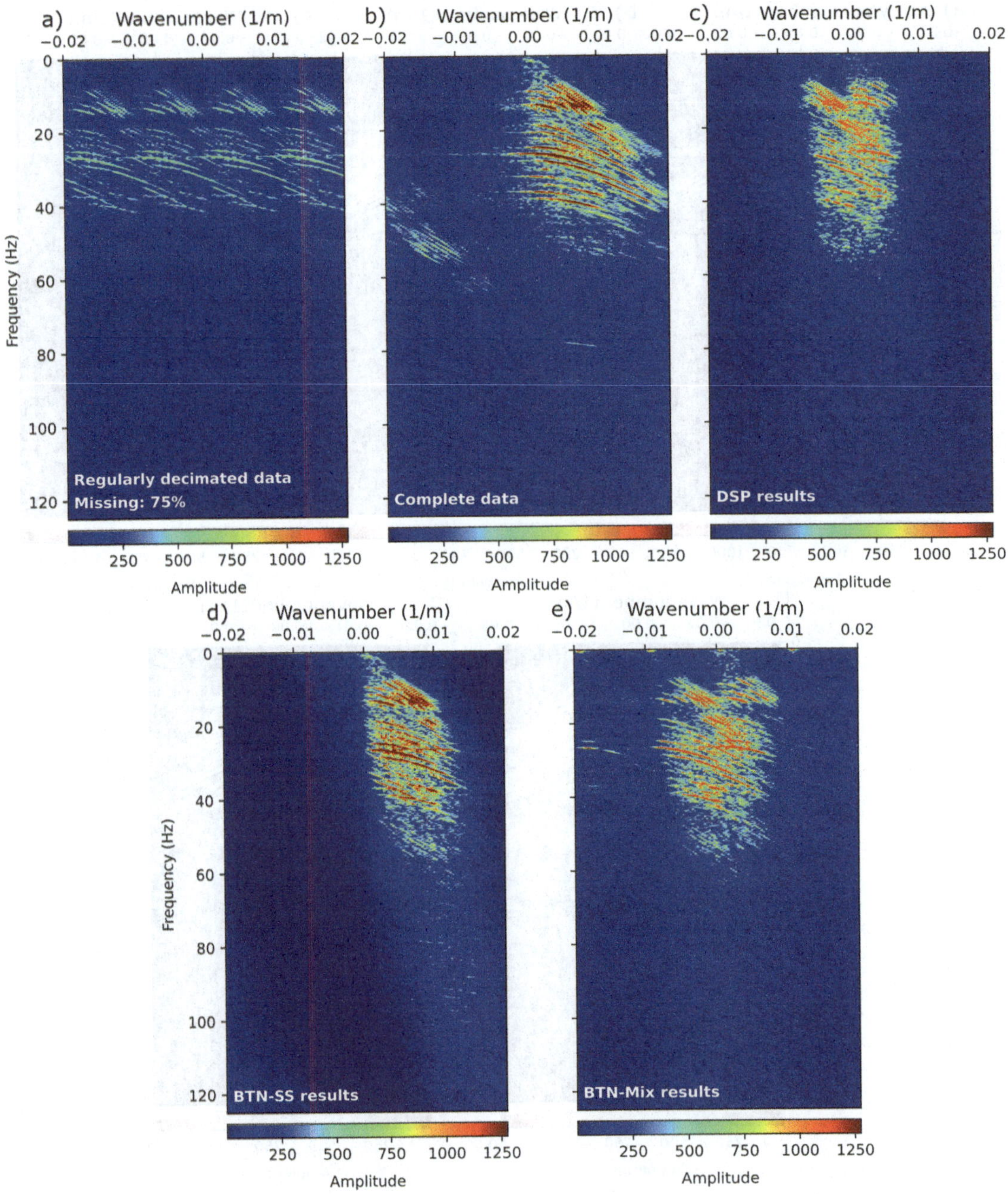

Fig. 4.34 F-k spectrum comparison for the regularly decimated Mobile AVO Viking Graben line 12 dataset with a missing ratio of 75%. (**a**) Decimated observation where the missing traces are inserted as 0s; (**b**) Complete seismic gathers; (**c**) Interpolated result by the DSP method; (**d**) Interpolated result by the BTN-SS method; (**e**) Interpolated result by the BTN-Mix method

features. Therefore, incorporating the spectrum mask in the f-k domain effectively regularizes the reconstructed components within a desired region, explaining the success of the BTN-SS method in Figs. 4.33d and 4.34d. The detailed trace comparison for the 75% decimation case is presented in Fig. 4.35, which further illustrates the superiority of the BTN-SS method.

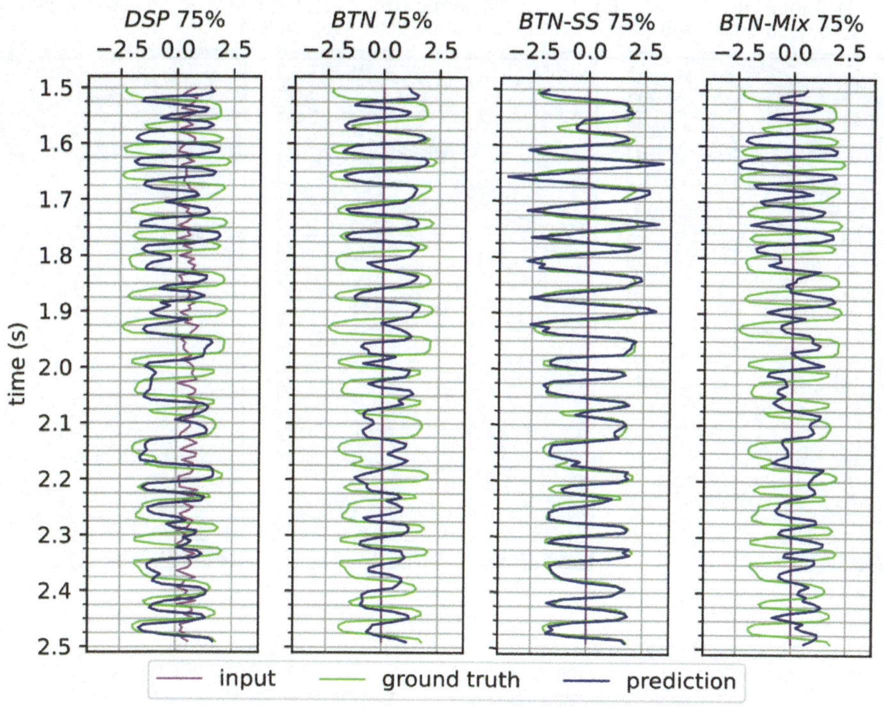

Fig. 4.35 Example traces of the Mobil AVO Viking Graben line 12 dataset with a regular missing ratio of 75%

4.4.4 Generalization Ability of BTN

The question of the BTN's generalization capability is indeed an intriguing aspect of our proposed method, particularly in the context of self-supervised seismic data interpolation. Typically, for such tasks, the network is trained directly on a target dataset and then applied to the same dataset for inference, meaning the training and inference data are identical. Our research has already showcased the BTN's ability to generalize from undecimated traces to decimated ones. However, another interesting aspect to explore is whether the BTN can generalize across different data volumes. In particular, we aim to ascertain if a BTN, self-trained on some decimated seismic gathers, can be effectively applied to a new, previously unseen data volume. This scenario closely resembles supervised interpolation tasks, with the key difference being that the training set in our case consists exclusively of decimated images.

In this experiment, we focus on interpolating the randomly decimated Mobile AVO Viking Graben line 12 dataset, applying a decimation ratio of 75%. To evaluate the BTN's adaptiveness across different shot gathers, the whole dataset is orderly partitioned into ten segments, or folds. And we reserve last fold for evaluating the network's performance. The volume of training data is then progressively increases from 10 to 90% as we gradually included one to nine of these folds. This strategy enable us to develop nine different models, each trained with varying amounts of shot gathers ranging from as little as 10% to as much as 90% of the total dataset. The variations in performance across these models are illustrated in Figs. 4.36 and 4.37. An observation from Fig. 4.36b is that with a limited volume of training data, some artifacts and noises are present, yet the overall performance remains satisfactory, with major seismic patterns accurately reconstructed. As the volume of training data increases, there is a noticeable improvement in performance, evidenced by the progression from Fig. 4.36b through 4.37e. Notably, the model trained with 9 folds closely approximates a fully self-supervised

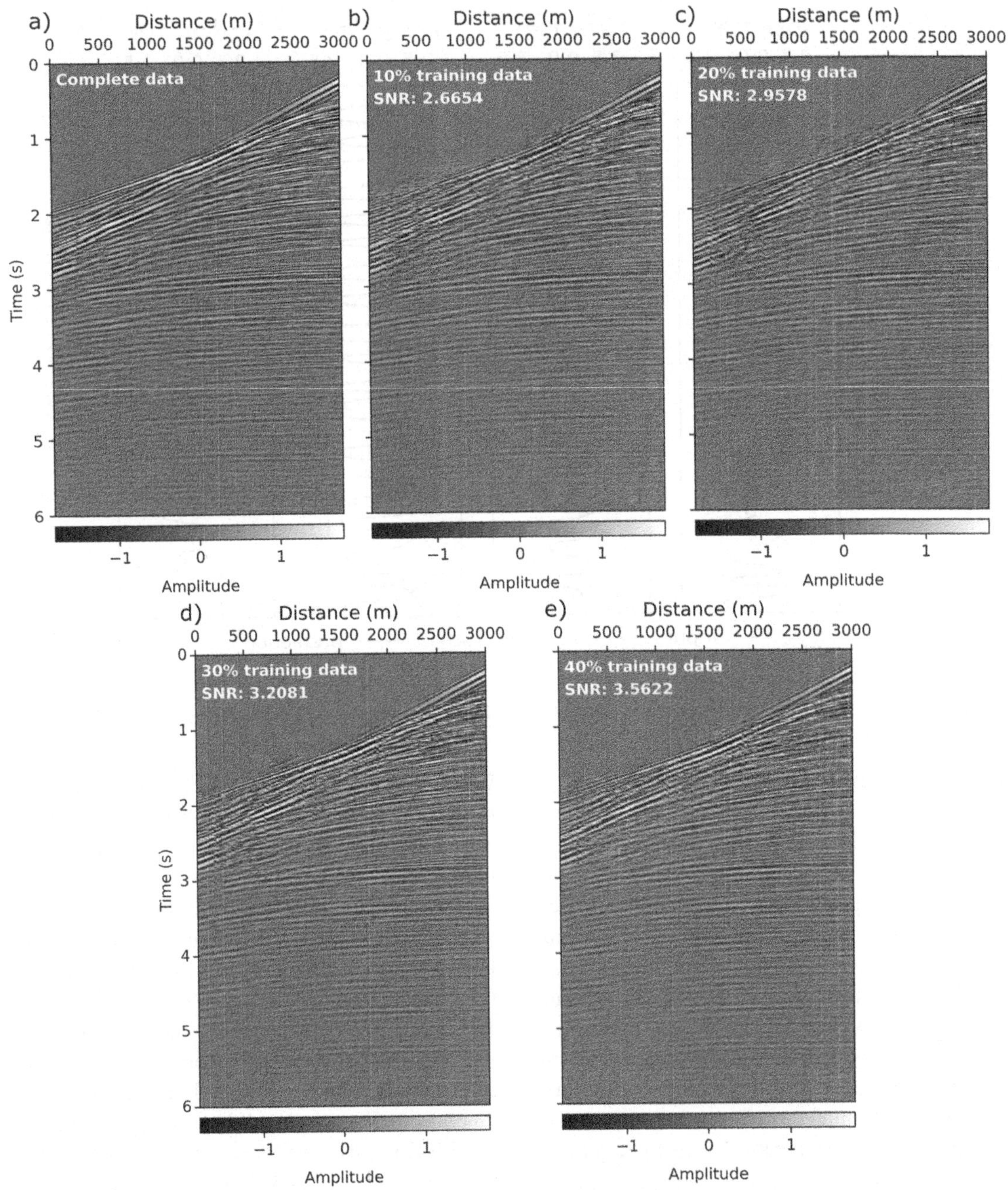

Fig. 4.36 Analyzing the generalization of the blind-trace network using the field dataset, which is randomly decimated with a missing ratio of 75%. (**a**) Complete sgather from the testing set; (**b**)-(**e**) Interpolated images from BTN when using partial of the training set ranging from 10% to 40%

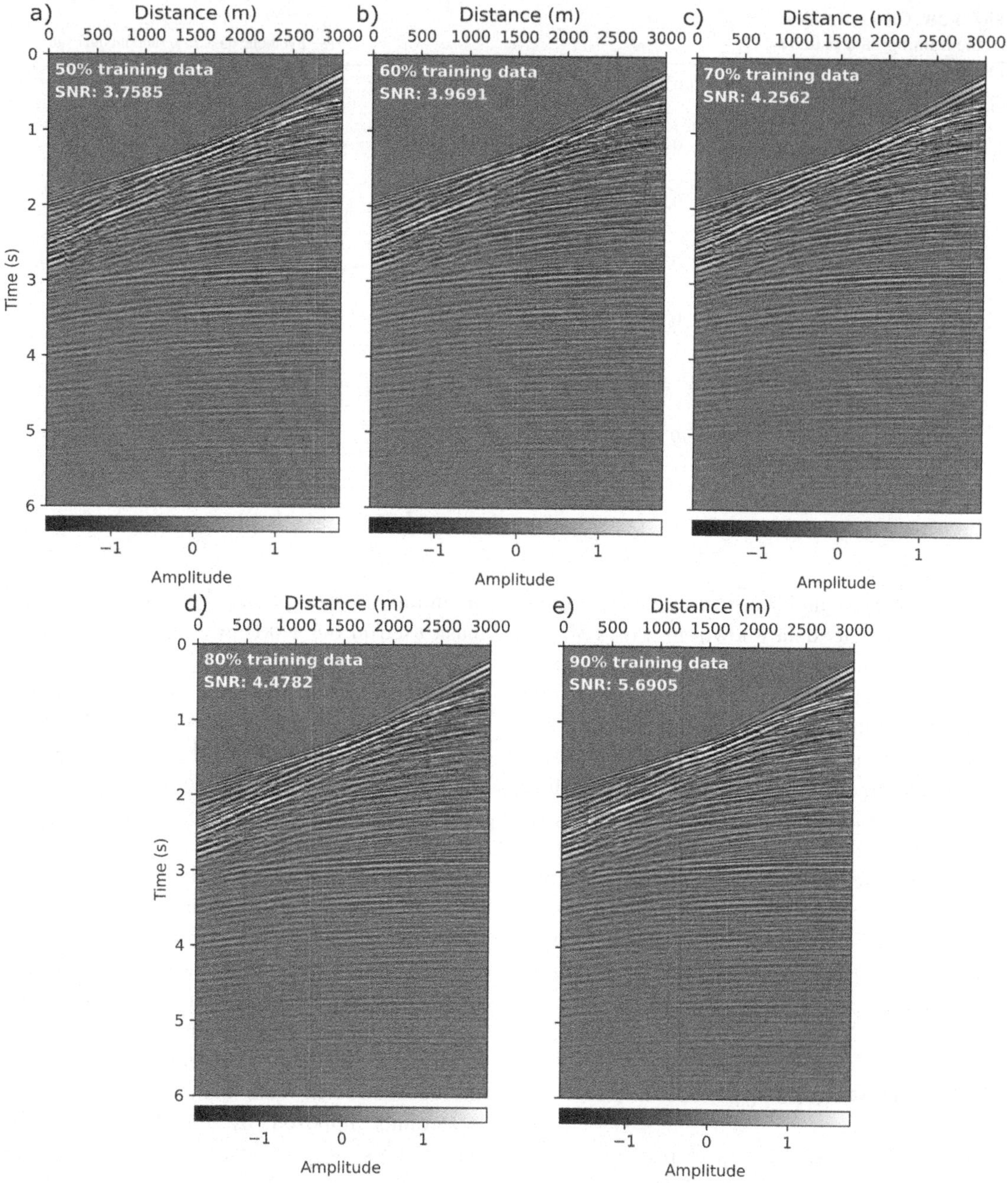

Fig. 4.37 Interpolated images from BTN when using partial of the training set ranging from 50% to 90%

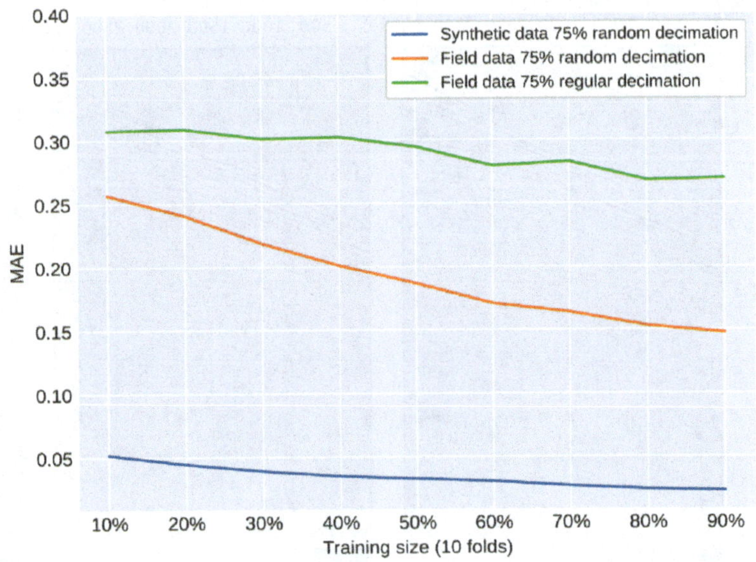

Fig. 4.38 Change of loss as the training data volume increases under different decimation scenarios

learning scenario and bears a strong resemblance to the complete data shown in Fig. 4.36a. A similar experiment is conducted on the SEAM-II unconventional dataset, also with a 75% regular decimation ratio. The trends observed in Fig. 4.38 which showcases the variation of losses relative to the volume of training data, echo the findings from the Mobile AVO Viking Graben line 12 dataset. These results underscore the robustness of our proposed method and suggest its potential in scenarios where training on a reduced amount of data is preferable, striking a balance between time and cost efficiency and minimal compromise in accuracy.

4.4.5 Impact of Suppression Mask

The implementation of a suppression mask is a pivotal aspect of the BTN-SS method, significantly influencing its performance.In our approach, we employ an automated mask generation method using the Hough transform. However, there are instances where this method may not yield an appropriate mask, particularly if the two boundaries are not distinctly detectable in the frequency-wavenumber (f-k) spectrum. In such cases, manual intervention is required, wherein a mask of any shape can be drawn based on intuitive assumptions. For the cone-shaped mask used in our work, we presuppose that the cone always originates at the top middle of the f-k spectrum, with the apex situated at the zero point. The two boundaries emanating from this apex delineate the sides of the dominant energy regions.

To explore the influence of f-k masks on interpolation results, we conduct experiments using the Mobile AVO Viking Graben line 12 dataset under a 75% regular decimation ratio. The findings of this investigation are presented in Figs. 4.39, 4.40, and 4.41. As illustrated in Fig. 4.9 we generated three distinct masks with varying boundary angles. Observations from the right column of Fig. 4.41 reveal that the reconstructed data adheres strictly to the constraints imposed by the f-k mask. This underscores the significant role of the f-k mask as a guiding constraint for the network during the reconstruction process. It's worth noting that the use of such cone-shaped assumptions is common in conventional seismic data processing and analysis algorithms. Additionally, experienced geologists typically do not find it challenging to identify a suitable f-k mask in most scenarios, which further validates the practicality of this approach in seismic data interpolation tasks.

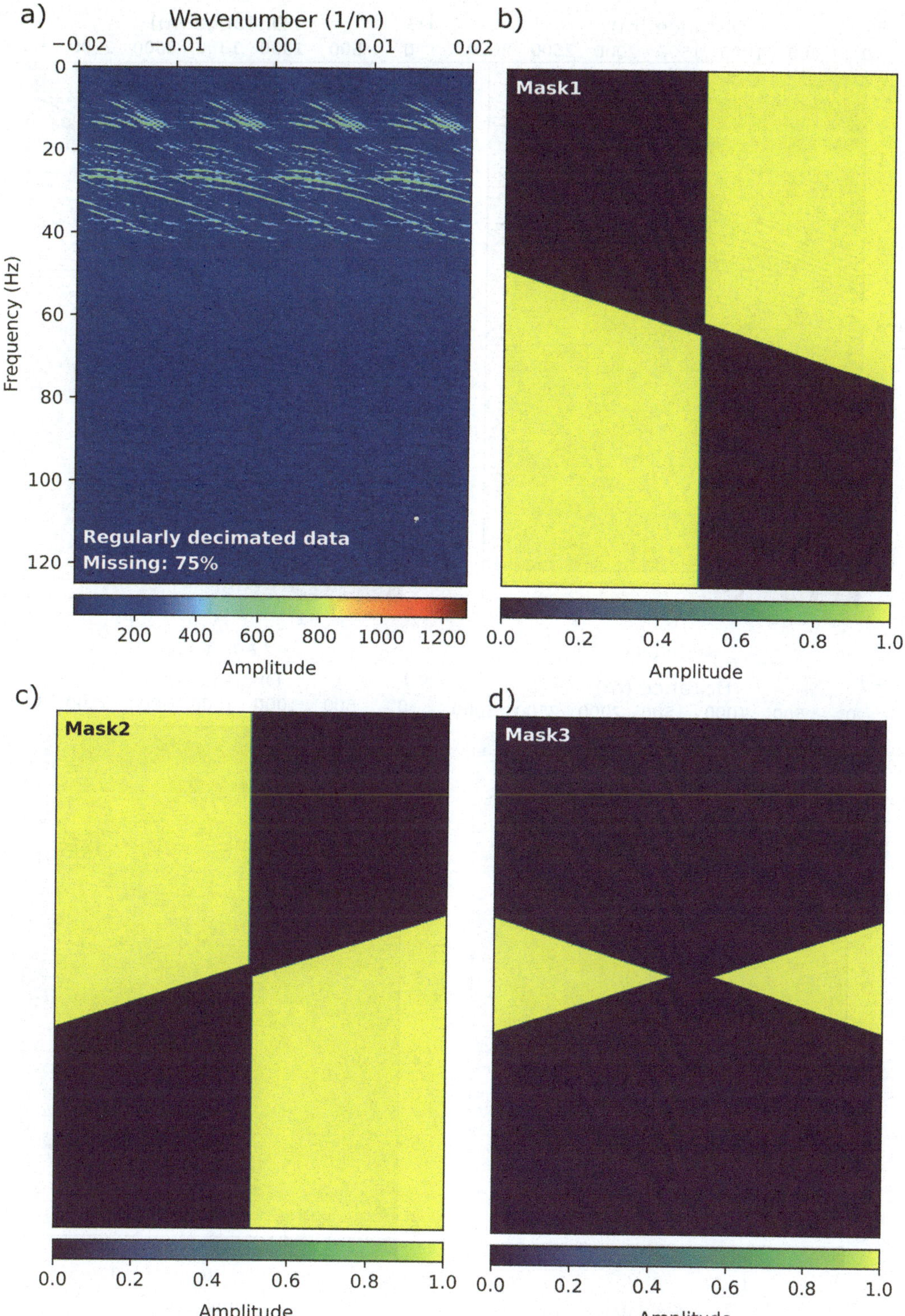

Fig. 4.39 Different mask selection. (**a**) F-k spectrum for the decimated target; (**b**)-(**d**) Three distinct mask generations

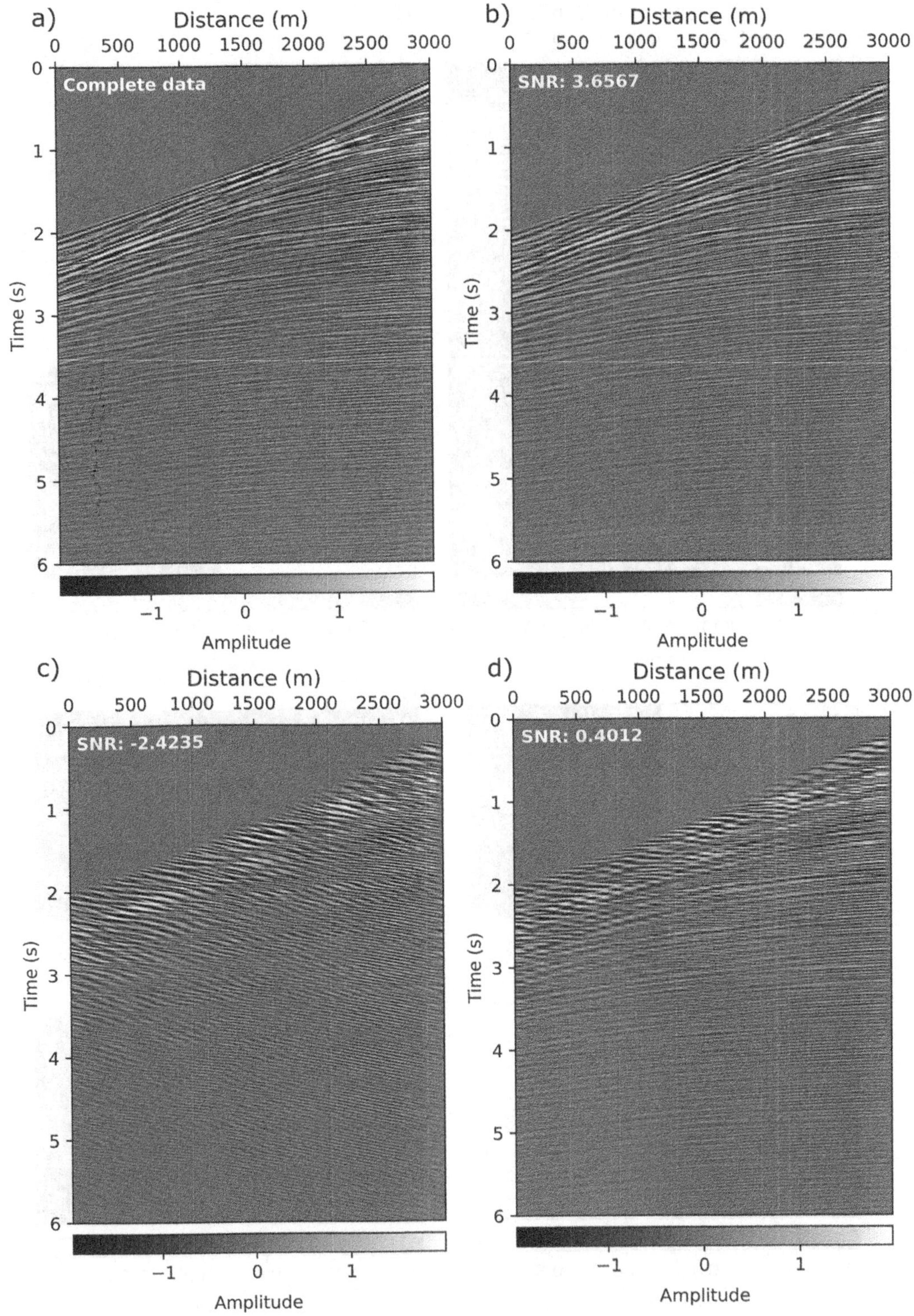

Fig. 4.40 Impact of the spectrum suppression mask on the field dataset with 75% regular decimation. (**a**) Complete image corresponding to Fig. 4.39 (a); (**b**)–(**d**) Interpolation results using different masks in Fig. 4.39 (**b**)–(**d**)

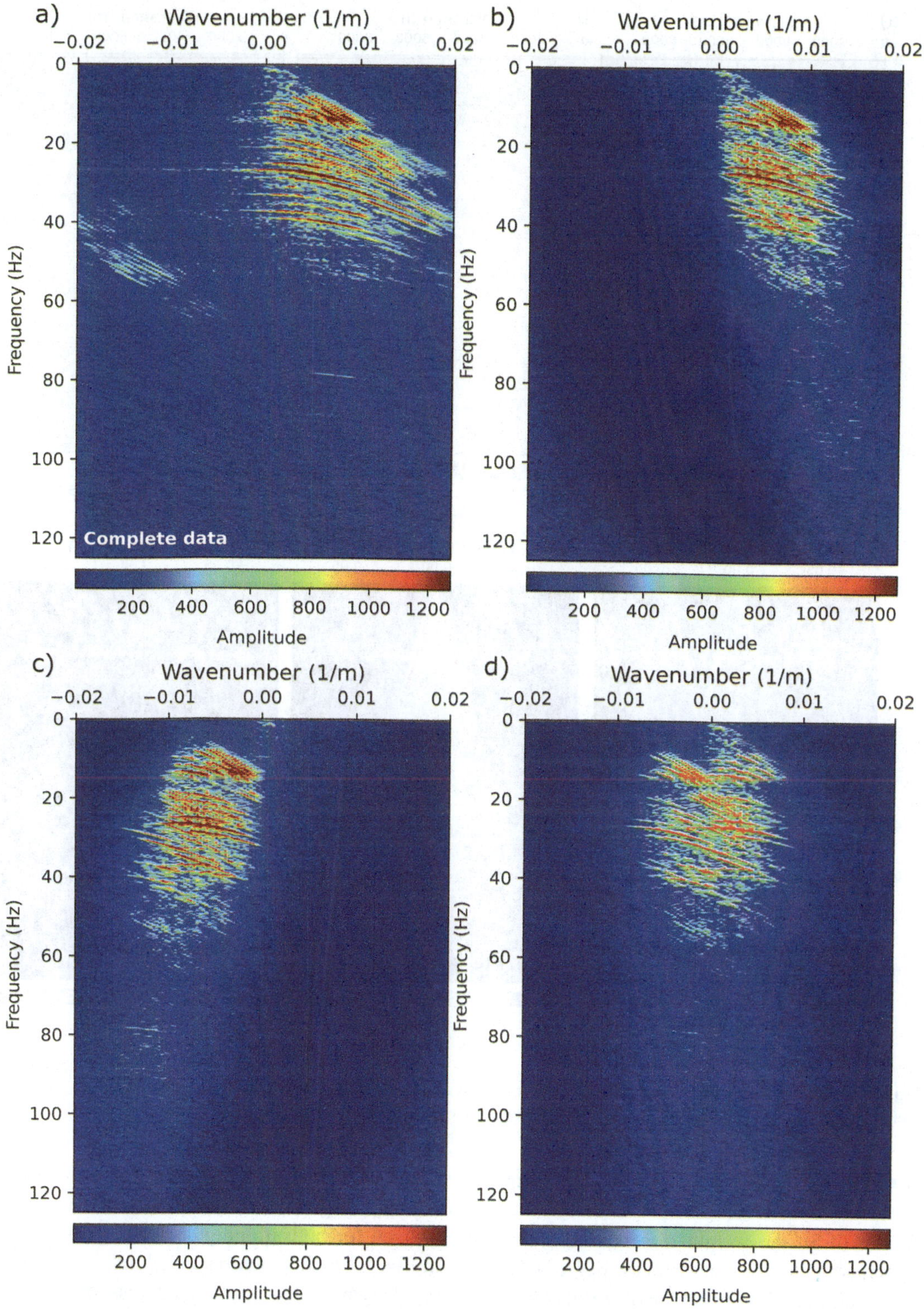

Fig. 4.41 Impact of the spectrum suppression mask on the field dataset with 75% regular decimation. (**a**) F-k spectrum of the complete image corresponding to Fig. 4.39 (a); (**b**)–(**d**) F-k spectra of the interpolation results using different masks in Fig. 4.39 (**b**)–(**d**)

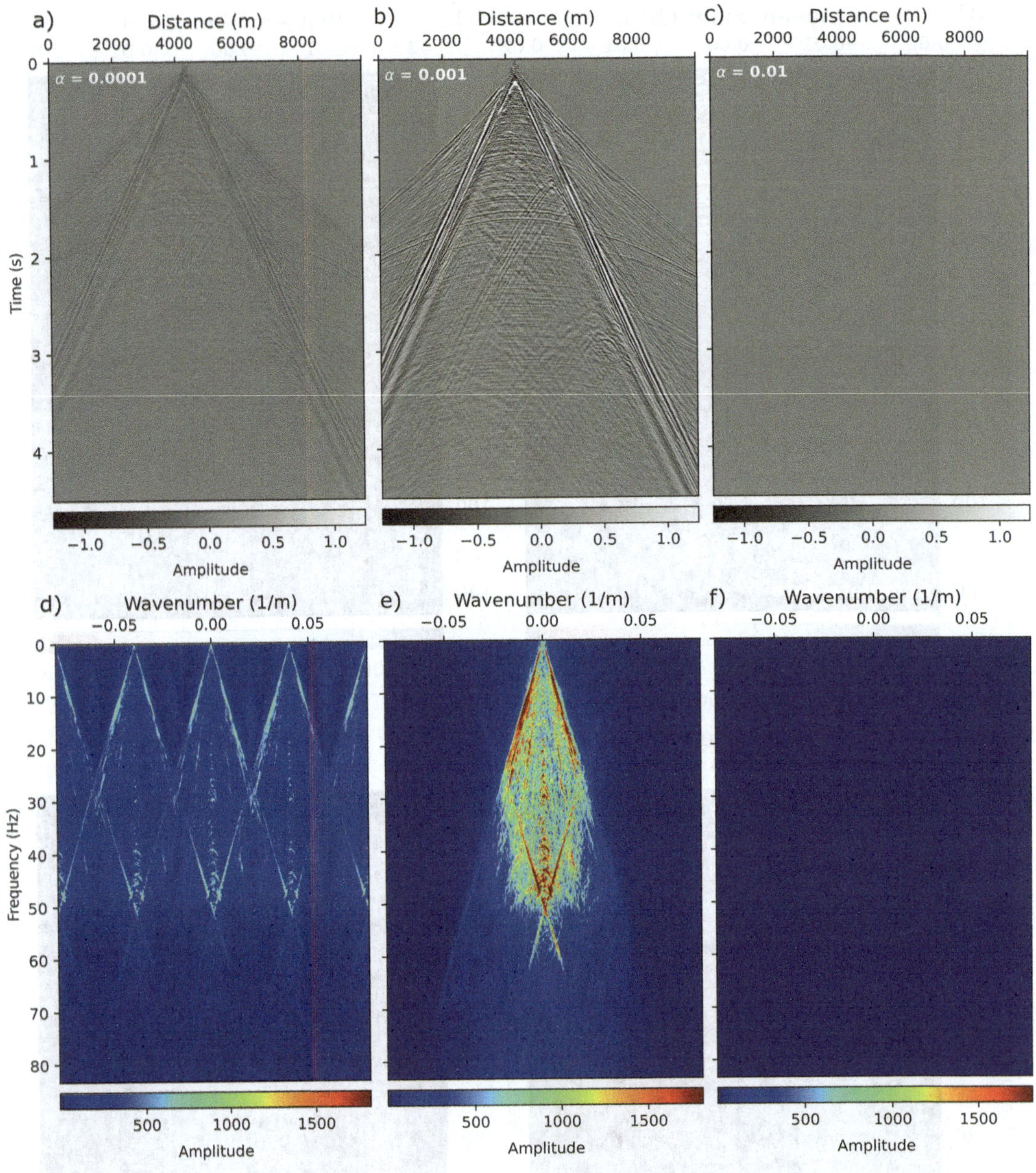

Fig. 4.42 The impact of the weight for regularization loss. (**a**)–(**c**) Visualization of the interpolation results when the weight parameter α increase from 0.0001 to 0.01, respectively; (**d**)–(**f**) Corresponding f-k spectra for (**a**)–(**c**)

4.4.6 Impact of Regularization Loss

In Eq. (4.5), the weight α serves as a crucial hyperparameter for managing the degree of regularization term. Figure 4.42 analyzes the effect of varying regularization loss through the selection of different $\alpha = 0.01, 0.001, 0.0001$ using the SEAMII–unconventional dataset. A single decimated image with a 75% decimation ratio is utilized for both training and testing. As shown in Fig. 4.39a, it is demonstrated that a very small regularization loss cannot address the aliasing problem and the result degrade to the naive BTN method. Conversely, when the regularization loss is excessively large, the network focuses on suppressing the unwanted components, resulting in predictions with entirely zero amplitudes, thereby achieving a very small regularization loss. Our experiments reveal that an α value of 0.001 is optimal for both the synthetic and field datasets employed in our work.

4.4.7 Discussing Advantages and Limitations of BTN

Common studies for the seismic data interpolation task with deep learning techniques assume that sufficient complete seismic gathers are always available for training the interpolation model in a supervised learning manner (Chai et al., 2020a). Such supervised learning scheme often suffers from the lack of appropriate training datasets and questioned generalization capability. By formatting a self-supervised learning strategy, our approach fully eliminates the need for the distributional concern between training and testing data and ensures the reliability of the trained networks. However, such a self-supervised learning scheme is not easy to be accomplished by a traditional convolutional neural network (CNN) such as the U-net. A crucial problem of identity mapping due to the ill-posedness nature of the interpolation problem is devastating to the learning process and leads to trivial solutions as shown in Figs. 4.12c and 4.14c. To address this problem, one can replace the decimated input with a random initialized noise for the network to prevent the direct processing of the undecimated true answers following the DSP method (Liu et al., 2021). However, the DSP method employing a generative model is often with slow convergence speed, and the results are often suboptimal due to the lack of seismic information in the input.

To utilize decimated gather as the input while preventing identity mapping, we introduce the blind-trace network (BTN) that employs a strict constraint for preventing the leakage of undecimated traces from the input to the prediction, which is instrumental to the effectiveness of our approach. Essentially, the BTN is supervised by the reconstruction of the undecimated traces during training, and subsequently applied to the missing traces during inference. The performance can be impacted by various aspects including decimation ratio, decimation scheme, the maximum and minimum sizes of the missing gaps, and so on. Therefore, the generalization concern can be a potential concern between the known and unknown traces. In random decimated scenarios, it is not a major concern since both the decimated and undecimated traces are randomly sampled with identical distribution in terms of both amplitudes and input decimation patterns. We have demonstrated that the BTN successfully accomplishes the self-supervised interpolation task, and outperforms the baseline methods both qualitatively and quantitatively.

However, this conclusion does not hold true for the regular seismic data interpolation task. In regular decimation scenarios where missing traces are evenly distributed in the seismic gather, such regularity of the missing patterns can cause significant distributional differences in terms of the input patterns, causing a severe overfitting problem for the BTN as presented in Fig. 4.18. Consequently, regularizations are indispensable. The DSP method addresses this issue by using a randomized input noise. And we proposed the spectrum suppression and the mix-training methods to incorporate priors for the training. All methods can effectively solve the overfitting problem as illustrated in Figs. 4.27, 4.28, 4.29, 4.30, 4.31 and 4.32. But the BTN-based methods have faster convergence speed and show higher accuracy in reconstructing complex patterns.

Another critical issue for regular seismic data interpolation is spectrum aliasing. In the absence of specific constraints, deep neural networks tend to yield predictions in the low-wavenumber regions, where the high-wavenumber components from the discretized spectra may interfere due to insufficient spatial sampling and the complex nature of the seismic data (as illustrated in the DSP results, Figs. 4.30c and 4.34c). To address this problem, the BTN-SS method provides explicit restrictions for the reconstructed energy to constrain it within a

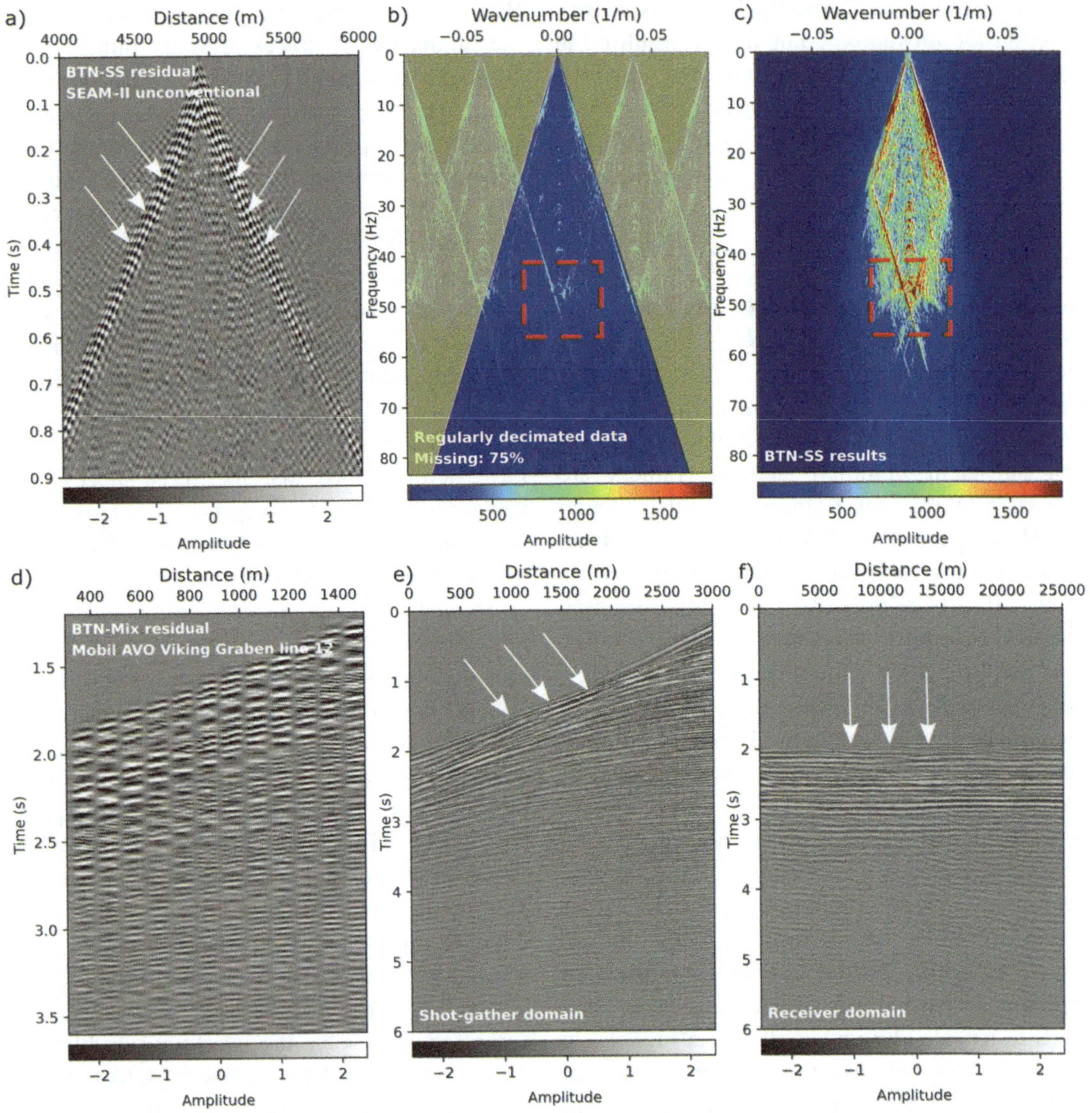

Fig. 4.43 Cases when the proposed BTN-SS and BTN-Mix approaches are challenged. (**a**)–(**c**) show that the BTN-SS method cannot handle the 75% regular decimated SEAM-II unconventional data, and (**d**)–(**f**) illustrate the BTN-Mix fails for the 75% regular decimated Mobile AVO Viking Graben line 12 dataset. (**a**) Residual between the reconstructed SEAM-II unconventional data and its corresponding ground truth in the 75% regular decimation case; (**b**), (**c**) The f-k spectrum of the corresponding decimated input and prediction, respectively; (**d**) Residual between the reconstructed SEAM-II unconventional data and its corresponding ground truth in the 75% regular decimation case; (**e**), (**f**) An example of the complete gather in the common shot domain and an exampled common receiver gather, respectively; when the source spacing is drastically different than the receiver spacing, resulting in different patterns between the training dataset and the auxiliary dataset

cone-shaped area, while the BTN-Mix method incorporates additional seismic priors using the auxiliary dataset. However, both methods exhibit certain drawbacks. As displayed in Fig. 4.43a–c, the BTN-SS method cannot handle the aliased components when the strong aliasing happens at the low-wavenumber region within the estimated f-k mask. This is inevitable for the BTN-SS method since the f-k mask only assumes the unaliased area outside of the mask. When the decimation ratio increases, it is not possible for the spectrum suppression method to eliminate the aliased energy in the low-wavenumber region. For the BTN-Mix problem, obtaining a suitable auxiliary dataset is not trivial in many scenarios. As demonstrated in Fig. 4.43d–f, although the common receiver gathers are always available, the difference in the trace spacing can also lead to completely different patterns in the seismic gathers. Addressing the aliasing issue is challenging. In the next chapter, we will extend the exploration of this critical issue for seismic data interpolation. Nonetheless, both methods introduced in this work can effectively address this problem in certain scenarios. And the performances of the proposed methods can always surpass the baseline method in both convergence speed and interpolation accuracy. In addition, the two proposed regularization methods can be flexibly combined with any other machine learning algorithms for seismic data processing, including the DSP approach.

4.5 Summary

In this chapter, we proposed a novel self-supervised deep-learning approach for the seismic data interpolation task. The critical identity mapping issue in the self-supervised seismic data interpolation problem is addressed by the blind-trace network (BTN). Employing the blind-trace constraint, the simple short-cut solution is strictly prevented, enforcing the network to learn the interpolation operation for a given decimated input. Our approach to our knowledge firstly uses the whole decimated gather as the network input compared to previous self-supervised learning schema (such as the DSP method). Besides, we explore the implementation of the BTN on both irregular and regular sampling scenarios.

The proposed approaches are evaluated with both synthetic and field datasets under high missing ratios, each demonstrating unique challenges. We explore the 75 and 87.5% missing ratios for irregular decimation cases, and 50 and 75% for the regular scenarios. For irregular decimated data, we demonstrate the BTN can successfully reconstruct the missing traces with the highest missing ratio of 87.5%, remarkably outperforming the baseline DSP method in both training speed and reconstruction quality. For regular decimated data facing the input pattern overfitting and spectrum aliasing issues, we proposed two efficient anti-aliasing techniques, namely BTN-SS and BTN-Mix. Either of them can successfully address the underlying problems and outperform the existing literature, delivering accurate interpolation results in various challenging scenarios.

References

Abma, R., Kabir, N.: 3D interpolation of irregular data with a POCS algorithm. Geophysics **71**(6), E91–E97 (2006)

Chai, X., Gu, H., Li, F., Duan, H., Hu, X., Lin, K.: Deep learning for irregularly and regularly missing data reconstruction. Sci. Rep. **10**(1), 1–18 (2020a)

Chai, X., Tang, G., Wang, S., Lin, K., Peng, R.: Deep learning for irregularly and regularly missing 3-D data reconstruction. IEEE Trans. Geosci. Remote Sens. (2020b)

Chang, D., Yang, W., Yong, X., Li, H.: Generative adversarial networks for seismic data interpolation. In: SEG 2018 Workshop: SEG Maximizing Asset Value Through Artificial Intelligence and Machine Learning, Beijing, China, 17–19 September 2018, pp. 40–43. Society of Exploration Geophysicists and the Chinese Geophysical Society, Beijing (2018)

Chang, D., Yang, W., Yong, X., Li, H.: Seismic data interpolation with conditional generative adversarial network in time and frequency domain. In: SEG Technical Program Expanded Abstracts 2019, pp. 2589–2593. Society of Exploration Geophysicists, Houston (2019)

Duda, R.O., Hart, P.E.: Use of the Hough transformation to detect lines and curves in pictures. Commun. ACM **15**(1), 11–15 (1972)

Fang, W., Fu, L., Zhang, M., Li, Z.: Seismic data interpolation based on U-net with texture loss. Geophysics **86**(1), V41–V54 (2021)

Jia, Y., Ma, J.: What can machine learning do for seismic data processing? An interpolation application. Geophysics **82**(3), V163–V177 (2017)

Kaur, H., Pham, N., Fomel, S.: Seismic data interpolation using CycleGAN. In: SEG Technical Program Expanded Abstracts 2019, pp. 2202–2206. Society of Exploration Geophysicists, Houston (2019)

Kaur, H., Pham, N., Fomel, S.: Seismic data interpolation using deep learning with generative adversarial networks. Geophys. Prospect. **69**(2), 307–326 (2021)

Keys, R.G., Foster, D.J.: A data set for evaluating and comparing seismic inversion methods. In: Comparison of Seismic Inversion Methods on a Single Real Data Set, pp. 1–12, Society of Exploration Geophysicists, Houston (1998)

Kong, F., Picetti, F., Lipari, V., Bestagini, P., Tubaro, S.: Deep prior-based seismic data interpolation via multi-res U-net. In: SEG International Exposition and Annual Meeting. OnePetro, Richardson (2020)

Liu, Q., Fu, L., Zhang, M.: Deep-seismic-prior-based reconstruction of seismic data using convolutional neural networks. Geophysics **86**(2), V131–V142 (2021)

Mandelli, S., Lipari, V., Bestagini, P., Tubaro, S.: Interpolation and denoising of seismic data using convolutional neural networks. arXiv:1901.07927 (2019)

Oropeza, V., Sacchi, M.: Simultaneous seismic data denoising and reconstruction via multichannel singular spectrum analysis. Geophysics **76**(3), V25–V32 (2011)

Pan, S., Chen, K., Chen, J., Qin, Z., Cui, Q., Li, J.: A partial convolution-based deep-learning network for seismic data regularization. Comput. Geosci. **145**, 104609 (2020)

Park, M.J., Jennings, J., Clapp, B., Biondi, B.: Seismic data interpolation using a POCS-guided deep image prior. In: SEG Technical Program Expanded Abstracts 2020, pp. 3154–3158. Society of Exploration Geophysicists, Houston (2020)

Regone, C., Stefani, J., Wang, P., Gerea, C., Gonzalez, G., Oristaglio, M.: Geologic model building in SEAM Phase II—Land seismic challenges. Lead. Edge **36**(9), 738–749 (2017)

Ronneberger, O., Fischer, P., Brox, T.: U-net: convolutional networks for biomedical image segmentation. In: International Conference on Medical Image Computing and Computer-Assisted Intervention, pp. 234–241. Springer, Berlin (2015)

Sen, S., Kainkaryam, S., Ong, C., Sharma, A.: Interpolation of regularly sampled prestack seismic data with self-supervised learning. In: SEG Technical Program Expanded Abstracts 2019, pp. 3974–3978. Society of Exploration Geophysicists, Houston (2019)

Spitz, S.: Seismic trace interpolation in the FX domain. Geophysics **56**(6), 785–794 (1991)

Ulyanov, D., Vedaldi, A., Lempitsky, V.: Deep image prior. In: Proceedings of the IEEE Conference on Computer Vision and Pattern Recognition, pp. 9446–9454 (2018)

Wang, B., Zhang, N., Lu, W., Wang, J.: Deep-learning-based seismic data interpolation: a preliminary result. Geophysics **84**(1), V11–V20 (2019)

Wang, Y., Wang, B., Tu, N., Geng, J.: Seismic trace interpolation for irregularly spatial sampled data usingconvolutional autoencoder. Geophysics **85**(2), V119–V130 (2020)

Yoon, D., Yeeh, Z., Byun, J.: Seismic data reconstruction using deep bidirectional long short-term memory with skip connections. IEEE Geosci. Remote Sens. Lett. **18**, 1298–1302 (2020)

Zwartjes, P., Sacchi, M.: Fourier reconstruction of nonuniformly sampled, aliased seismic data. Geophysics **72**(1), V21–V32 (2007)

5 Self-Supervised Learning for Anti-Aliased Seismic Data Interpolation Using Dip Information

5.1 Introduction

5.1.1 Aliasing Issue for Seismic Data Interpolation

Aliasing introduces a significant and enduring challenge in the field of seismic data interpolation. This issue arises when the spatial sampling rate is insufficient to accurately capture the intricate spatial variations present in the underlying seismic wavefield. As a consequence, high-frequency components in the seismic signal may "alias" and present as low-frequency components, resulting in the introduction of artifacts and notable distortions within the seismic data. The extent and severity of the aliasing issue are influenced by factors such as the spatial sampling rate and the specific underlying geological setting, adding to its unpredictability and, in many cases, inevitability for seismic data processing and analysis.

Addressing the aliasing issue in seismic data interpolation has been a long-standing and complex challenge. Gülünay (2003) approached this challenge by proposing that dominant slopes maintain continuity in the frequency-wavenumber (f-k) spectrum of a seismic gather, spanning both low-frequency, low-wavenumber and high-frequency, high-wavenumber components. This insight enabled the creation of a mask in the f-k domain derived from low-pass filtered elements, which then guided the reconstruction of high-energy ratios, thereby effectively reducing aliasing. Similarly, Naghizadeh and Sacchi (2010) introduced a method to hierarchically filter aliased components using a manually selected threshold in the curvelet domain. In another approach, Naghizadeh (2012) utilized a masking method constrained by dominant dips in the f-k domain to regulate the least square solution for noise reduction and seismic data interpolation. Curry (2010) employed a Fourier-radial adaptive thresholding approach to the f-k spectrum of seismic data to ensure successful separation of aliased energy. Taking advantage of the cone-concentrated assumption of seismic data in the f-k domain, Zhang et al. (2020) proposed an f-k mask constraint to regularize the POCS (Abma and Kabir, 2006) iterations in the curvelet domain. Gan et al. (2015) incorporated a seislet thresholding constraint into the POCS iterations to prevent aliasing for regularly sampled seismic data interpolation. These methods critically relied on exploiting the sparsity assumption of unaliased seismic components in various transformed domains and integrated such regularizations with traditional algorithms. While conventional anti-aliasing methods in seismic data interpolation have yielded promising results, they often hinge on thresholding or masking operations, making the determination of the optimal hyperparameters a non-trivial task. This challenge is exacerbated in complex scenarios where the seismic data is heavily aliased, potentially leading to suboptimal results.

S. Wang et al., *Deep Learning for Seismic Data Enhancement and Representation*, Advances in Oil and Gas Exploration & Production, https://doi.org/10.1007/978-3-031-75745-7_5

The rapid evolution of deep learning technologies has also spurred notable progress in anti-aliasing for seismic data interpolation. A deep convolutional neural network (CNN) was utilized by Wang et al. (2019) to interpolate regularly sampled seismic data without adding explicit anti-aliasing constraints to the model. Fang et al. (2021) combined the prediction-error filter (PEF) with deep learning, enabling the neural network to inherently learn local dips and reconstruct unaliased seismic traces. Generative approaches have also been explored to distinguish between aliased and unaliased seismic data. A conditional Wasserstein Generative Adversarial Network (WGAN) was adopted by Wang et al. (2019) for regular seismic data interpolation. Additionally, a deep-prior-based anti-aliased seismic data interpolation was proposed by Picetti et al. (2021), in which a low-pass version of the interpolated data was used to compute the slope and a directional Laplacian operator was utilized to constrain the dips of reconstructed patterns. These developments underscore the potential of deep learning in addressing the aliasing challenge in seismic data interpolation. However, the aliasing issue has not yet received significant attention, and challenges persist in the aspects of model adaptiveness, training strategy design, and computational costs.

5.1.2 Challenges for Anti-Aliased Regular Interpolation by Self-Supervised Learning

The aliasing problem can occur in both regularly and irregularly sampled seismic gathers, resulting in overlapped f-k spectra where the low-frequency, low-wavenumber components interfere with the high-frequency, high-wavenumber components. However, regularly sampled gathers are often more prone to aliasing than irregular ones when subjected to the same decimation ratio, typically due to the higher maximum sampling rate in irregularly sampled seismic gathers resulting from the randomness of the decimation patterns. Consequently, devising an anti-aliasing deep learning algorithm for regular seismic data interpolation presents a greater challenge compared to irregular scenarios. Furthermore, developing a self-supervised solution to the problem adds another layer of complexity. Two major obstacles for regular seismic data interpolation under the self-supervised learning scheme are identified in Sect. 4.4, namely: (1) The variance between training and testing due to the consistent decimation patterns, and (2) The resulting aliased f-k spectrum from an insufficient sampling ratio. Effectively addressing these two obstacles is essential for the successful development of a robust self-supervised learning scheme for regularly sampled seismic data interpolation.

Despite being termed "self-supervised", the self-supervised seismic data interpolation strategy still features a notable disparity between its training and testing phases, as described in Chap. 4. In the training phase, the network is designed to learn the reconstruction of undecimated traces, a process guided by the loss function (Eq. (4.2))

$$L_\theta(\mathbf{m}) = \|\mathbf{x} - \sigma_{\mathbf{m}}(f_\theta(\mathbf{x}))\| \ , \tag{5.1}$$

meanwhile, with no supervision applied to the decimated traces regarding the interpolation loss (Eq. (4.3))

$$L_\theta(\widetilde{\mathbf{m}}) = \|\sigma_{\widetilde{\mathbf{m}}}(\mathbf{y}) - \sigma_{\widetilde{\mathbf{m}}}(f_\theta(\mathbf{x}))\| \ . \tag{5.2}$$

Therefore, the distinction in the distribution of training and testing input can significantly affect the performance of a trained neural network in this self-supervised learning scheme. Specifically, the local input decimation patterns between the decimated and undecimated traces play a crucial role in the effectiveness of self-supervised interpolation algorithms, especially for interpolating regularly sampled seismic gathers. To demonstrate this phenomenon in detail, we use a 50% decimation example as illustrated in Fig. 5.1. In the 50% random sampling case Fig. 5.1a, the local input decimation patterns for predicting the underlying central trace, the undecimated traces during training and decimated traces in the inference stage, are both random and

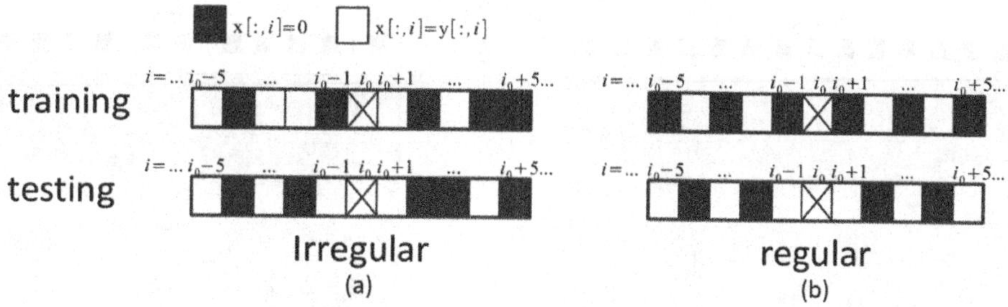

Fig. 5.1 Difference of the local decimation patterns between irregular and regular interpolation. Each row illustrates the surrounding pattern of a trace that needs to be reconstructed at position i_0, highlighted with a crossing pattern in the middle. Following the formulation of Eq. (4.1), each rectangle indicates a single trace of the seismic gather, with black rectangles indicating missing traces and white rectangles representing preserved traces. (**a**) Local decimation patterns during training ($\mathbf{x}[:, i_0] = \mathbf{y}[:, i_0]$) and testing ($\mathbf{x}[:, i_0] = 0$) for randomly sampled seismic data; (**b**) Decimation patterns for regular seismic data interpolation scenarios

share the same random distribution across the whole gather. However, in the regular sampled case under the same decimation ratio illustrated in Fig. 5.1b, the input patterns are fixed for all undecimated traces and opposite to all decimated traces. This notable discrepancy in decimation patterns between undecimated and decimated traces elucidates the failure of the blind-trace network (BTN) shown in Fig. 4.18.

The overfitting issue for regularly sampled seismic data interpolation is inevitable without adding constraints during the training phase, such as the BTN-SS method described in Sect. 4.4.1, or the introduction of randomness, as demonstrated by DSP and BTN-Mix in Sect. 4.4.2. However, as illustrated in Fig. 4.43, both strategies discussed in Sect. 4.4 face limitations in addressing the aliasing challenge. Although the BTN-SS approach constrains the reconstructed components within a cone-shaped area defined by the steepest events, it does not guarantee the removal of aliased energy within the narrow f-k mask. On the other hand, the BTN-Mix method presents a viable alternative by introducing randomness and incorporating dip information from the auxiliary training dataset. However, obtaining an optimal auxiliary training dataset is not always feasible. Additionally, the deep-prior-based approach, which employs a random input vector to introduce randomness, faces difficulties in managing cases with severe aliasing, as elaborated in Sect. 4.4.

In this chapter, we are dedicated overcoming the challenges associated with the self-supervised interpolation of regularly sampled seismic data. Our approach involves developing a randomized mixed training workflow. This workflow is designed to introduce input randomness for both decimated and undecimated traces, thereby effectively countering the issue of decimation pattern overfitting. Additionally, we innovatively introduce a dip-informed neural network tailored to incorporate dip information into its architecture, directly addressing the aliasing problem. Through comprehensive synthetic and field case studies, we demonstrate the enhanced performance of our proposed methods compared to existing techniques documented in the literature.

5.2 Anti-Aliasing Seismic Data Interpolation with a Dip-Informed Neural Network

5.2.1 Introducing Randomness for Regular Seismic Data Interpolation

As discussed in Sect. 5.1.2, introducing randomness during the training process can reduce the variance in local missing patterns between known and unknown traces, thereby alleviating the overfitting issue. Building upon the mix-training concept introduced in Sect. 4.4.2, we propose a novel methodology that generates the auxiliary dataset

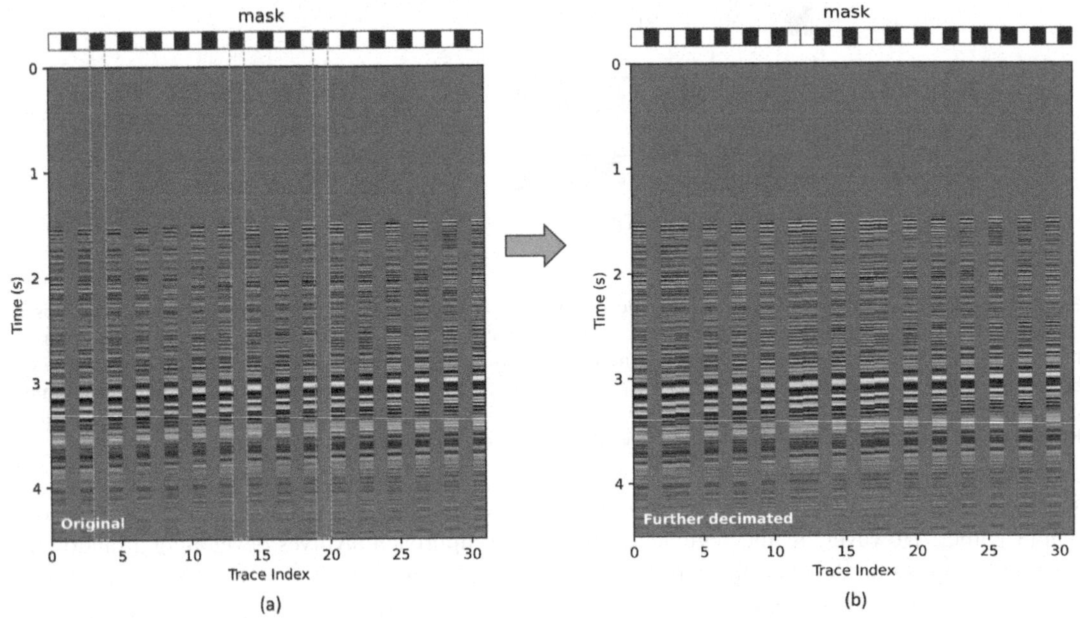

Fig. 5.2 Visualization of the further random decimation on the target dataset with an original sampling ratio of 50%. (**a**) A local segment of the target data that has undergone 50% regular sampling. The black and white stripe at the top indicates the missing patterns specified in Fig. 5.1. The traces marked by blue rectangles illustrate the traces to be further removed by a new mask $\mathbf{M}_0$ at this step. (**b**) The resulting segment after the masking process, where the remaining traces are concatenated to form a new random decimation pattern

directly from the original input gathers, instead of sourcing it from other domains or fields. The procedure for creating this dataset is depicted in Fig. 5.2. The process starts with the original, regularly sampled input, denoted as $\mathbf{x}$, in which unknown traces are filled with zeros. A subsequent decimation mask $\mathbf{M}_0$ is then applied, randomly eliminating the zero traces according to a predetermined decimation ratio, which results in an input of reduced size, denoted as:

$$\mathbf{x}' = \mathbf{M}_0\mathbf{x}. \tag{5.3}$$

The initial decimation mask $\mathbf{m}$ is concurrently processed by the same mask as

$$\mathbf{m}' = \mathbf{M}_0\mathbf{m}. \tag{5.4}$$

It should be emphasized that while the additional random masking procedure alters the geological configuration of the traces, this is not a cause for concern in our context since this alternation is only applied to the auxiliary training dataset. The correct spacing information is preserved by the original dataset, which will be also used for training and inference phases. The introduction of randomized decimation patterns in this dataset plays a crucial role in mitigating the issue of overfitting. Consequently, this approach significantly improves both the reliability and robustness of the model developed through this training process.

5.2.2 Dip-Informed Neural Network (DINN)

Addressing the aliasing problem for seismic data interpolation posses a significant challenge. Owing to the underdetermined nature of the interpolation problem, self-supervised neural networks often tend to generate trivial solutions where high-wavenumber energy is aliased into the low-wavenumber region. This phenomenon is evident in the f-k interpretation shown in Figs. 4.34c–e and 4.43c, highlighting the need for appropriate constraints and accurate guidance to prevent undesired aliasing in the network's predictions.

Conventional methods for preventing aliasing have emphasized the significance of dip information, and incorporated it into various algorithms (Gülünay, 2003; Naghizadeh, 2012; Gan et al., 2015). For deep learning approaches, Picetti et al. (2021) firstly incorporated dip information as a loss term during network training, aiming to constrain the slopes of predicted events. However, the direct incorporation of explicit dip information into the network architecture itself remains unexplored in the literature. Our research addresses this gap by introducing a dip-informed neural network (DINN), which directly harnesses the dip information of seismic gathers. This network aims to guide the reconstruction of events along local slopes, offering a novel approach in the field.

5.2.2.1 Dip-Informed Deformable Convolution (DIDC)

Fomel and Liu (2010) introduced the seislet transform which allows the kernels to be shifted regarding local dips in the process of discrete wavelet transform (DWT), resulting a sparser representation for seismic data compared to traditional DWT. Drawing inspiration from this dip-informed shifting concept in the Seislet transform, we propose the dip-informed deformable convolution (DIDC) layers to replace traditional convolutional layers in neural networks. Previously, Dai et al. (2017) had proposed a deformable convolution technique, by which an offset is imposed to disturbs the sampling grid of the convolutional kernel, enabling a more adaptable receptive field for different objects in the network input. In our work, we generate a vertical offset for all convolutional kernels based on the dip information. This adjustment ensures that computations align with the slopes of local seismic events, effectively addressing the aliasing issue.

The dip-informed deformable convolution is depicted in Fig. 5.3. Specifically, we define the sampling grid of a (3×3) convolution kernel as

$$\mathbf{P} = \{\mathbf{p}_{ij}\}_{i=0,j=0}^{2,2} = \begin{pmatrix} (-1,-1) & (-1,0) & (-1,1) \\ (0,-1) & (0,0) & (0,1) \\ (1,-1) & (1,0) & (1,1) \end{pmatrix}, \tag{5.5}$$

where the central point is denoted as $\mathbf{p}_{11} = (0,0)$. Therefore, for a location $\mathbf{p}_0 = (p_x, p_y)$, the traditional convolution is defined as

$$\mathbf{y}(\mathbf{p}_0) = \sum_i \sum_j \mathbf{w}(\mathbf{p}_{ij}) \cdot \mathbf{x}(\mathbf{p}_0 + \mathbf{p}_{ij}), \tag{5.6}$$

where $\mathbf{x}$ and $\mathbf{y}$ are the input and output, respectively.

To enable the dip-guided computation, a vertical offset $\Delta\mathbf{P}$ is derived from the dip map and applied to the convolutional kernel. With $\boldsymbol{\theta}$ representing the angle of the local event at $\mathbf{p}_0$, an local offset grid can be obtained as

$$\Delta\mathbf{P} = \{\Delta\mathbf{p}_{ij}\}_{i=0,j=0}^{2,2} = \begin{pmatrix} (-\tan\boldsymbol{\theta},0) & (0,0) & (\tan\boldsymbol{\theta},0) \\ (-\tan\boldsymbol{\theta},0) & (0,0) & (\tan\boldsymbol{\theta},0) \\ (-\tan\boldsymbol{\theta},0) & (0,0) & (\tan\boldsymbol{\theta},0) \end{pmatrix}, \tag{5.7}$$

thus the convolution in Eq. 5.6 is reformed as

$$\mathbf{y}(\mathbf{p}_0) = \sum_i \sum_j \mathbf{w}(\mathbf{p}_{ij}) \cdot \mathbf{x}(\mathbf{p}_0 + \mathbf{p}_{ij} + \Delta\mathbf{p}_{ij}). \tag{5.8}$$

The horizontal offsets within $\Delta\mathbf{P}$ at all indexing positions are set to zeros, indicating no lateral shifting throughout the computation. Furthermore, all vertical offsets within the same column are assigned identical values. The sampling at $\mathbf{x}(\mathbf{p}_0 + \mathbf{p}_{ij} + \Delta\mathbf{p}_{ij})$ is achieved by bilinear interpolation as described in Dai et al. (2017). In this work, the local dip map $\boldsymbol{\theta}$ is computed through plane wave destruction (PWD) (Fomel, 2002, 2006). However, it could also be manually assigned or derived using alternative techniques. The derived offset map shares the same size of the output in each layer and is composed of 18 channels, representing the horizontal and vertical offsets for all indexes in a (3×3) convolution kernel (Fig. 5.4).

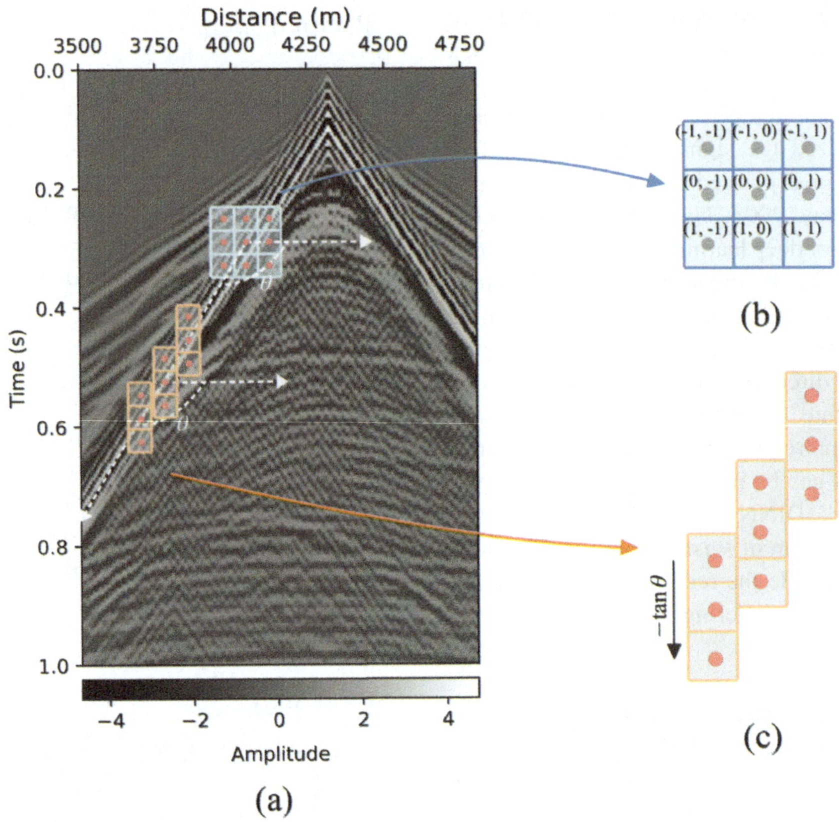

Fig. 5.3 Depict of dip-informed deformable convolution (DIDC). (**a**) Illustration of the deformable convolution concept for a local event with angle θ. (**b**) The rigid grid of a traditional convolution filter. (**c**) With dip information, the DIDC enables adaptive convolution kernels to align with the local event with a column-wise shift of $\tan\theta$

5.2.2.2 Network Structure

The architecture of the dip-informed neural network (DINN) is illustrated in Fig. 5.4. The DINN adapts the general framework of the U-net (Ronneberger et al., 2015) but introduces significant modifications to its convolutional layers. Specifically, each layer in the original U-net is replaced by a deformable residual block, as shown in Fig. 5.5a. This block is a variation of the standard residual block (He et al., 2016), wherein traditional convolutional layers are replaced with their deformable counterparts, i.e. the DIDC.

The DINN comprises four down-sampling and up-sampling stages. Critically, all traditional convolutional-like layers, including max-pooling and average-pooling layers must be avoided since it could potentially compromise the dip constraint. Therefore, we employ a naive down-sampling layer to remove every other rows and columns to reduce the size of each dimension by half, which is illustrated in Fig. 5.5b. Similarly, we avoid convolution in up-sampling layers by simply padding zero rows and columns to double the width and height of the feature maps, as depicted in Fig. 5.5c. The dip map is down-sampled after each contracting block using the same down-sampling approach to match the size of the feature maps and provide consistent offset maps for all convolutional layers, indicating the assumption of invariant slopes throughout the whole computations.

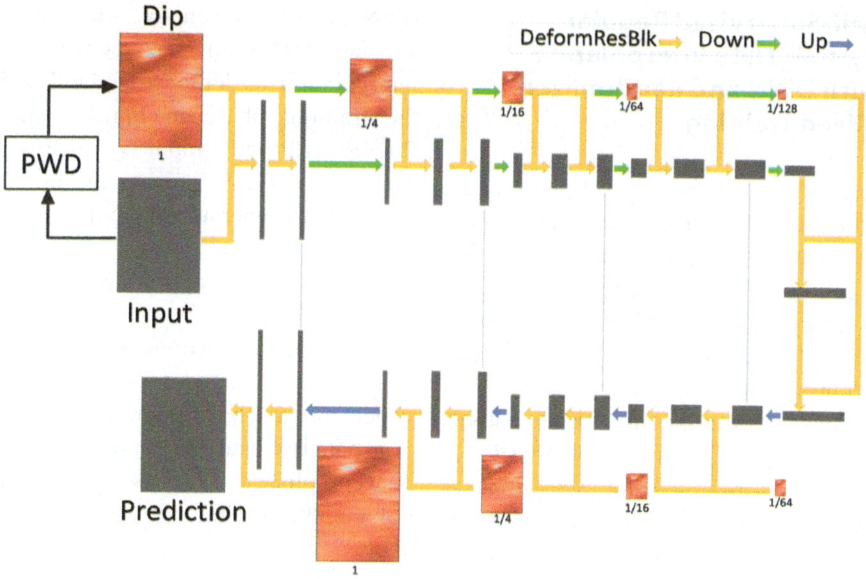

Fig. 5.4 Structure of the dip-informed neural network (DINN). Given a seismic image as input, a rough deep map is computed. in this work, we use the plane-wave destruction (PWD) (Fomel, 2002) to align with the formulation of Seislet Transform, but it can be potentially obtained by other methods. For each point, a local offset $\Delta\mathbf{P}$ is calculated from the dip map and applied to the kernels of deformable convolution

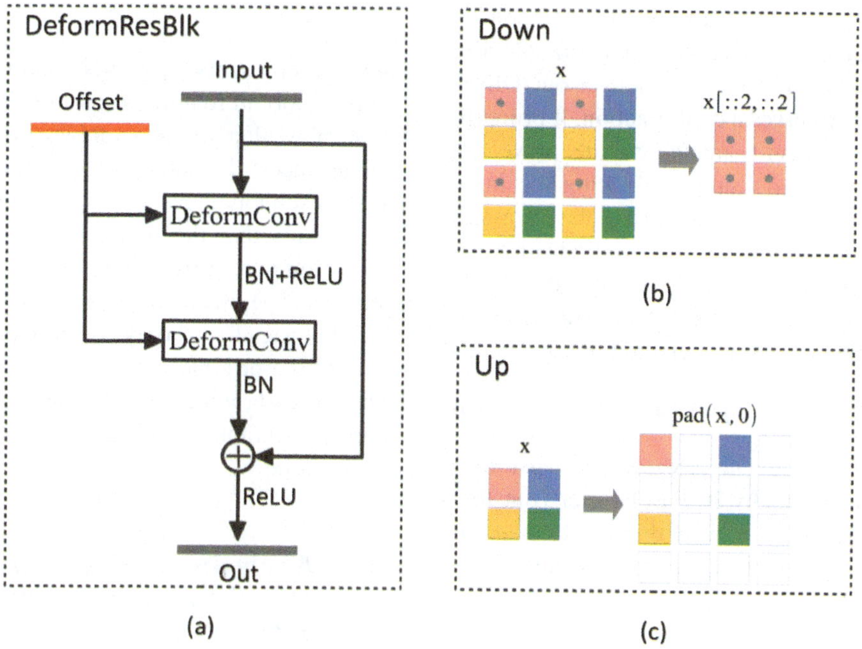

Fig. 5.5 Components of the DINN in Fig. 5.4. (**a**) Illustration of the deformable residual block that replaces the original convolutional layer in the U-Net structure. It is composed of two dip-informed deformable convolutional layers with the skip connections. The offset is derived from the dip map. (**b**)–(**c**) Down-sampling and up-sampling layers

5.2.3 Self-Supervised Regular Seismic Data Interpolation with DINN and Randomized Mixed Training

To successfully develop a self-supervised learning scheme for the seismic data interpolation task, the identity mapping issue must be addressed primarily, as detailed in Sect. 4.2.1. Identity mapping leads to a trivial solution where the network's prediction merely replicates the input, leaving the decimated traces unchanged due to data leakage from the input to the output. Methods based on the blind-trace network (BTN) implement a blind-trace constraint to conceal the correct answer from the input for each trace. Deep-prior-based methods, on the other hand, opt to initialize the inputs with random vectors. Diverging from these approaches, the proposed DINN-based method utilizes a low-pass filter in the frequency-wavenumber (f-k) domain to initially remove aliased components in the high-frequency and high-wavenumber components of an input gather. Consequently. the adoption of low-pass-filtered initialization in our DINN-based method serves a dual function. It primarily addresses the identity mapping issue, preventing the network from simply replicating the input. Additionally, this technique offers the advantage of providing a general dip initialization for the network, which contributes to reducing the aliasing problem to some extent. The details of this filtering process are depicted in Fig. 5.6, and denoted as

$$\mathbf{z} = \mathcal{F}^{-1}\mathbf{L}\mathcal{F}\mathbf{x}, \tag{5.9}$$

where the decimated input $\mathbf{x}$ is firstly transformed into the f-k domain by 2D Fast Fourier Transform (FFT), represented as $\mathcal{F}$. Within this domain, the spectrum is then modified using an assigned low-pass filter $\mathbf{L}$, and the filtered spectrum is subsequently converted back to the time-space domain using the Inverse FFT (IFFT) operator $\mathcal{F}^{-}1$. The outcome of this sequence, denoted as $\mathbf{z}$, is the low-pass filtered (LPF-ed) input.

By integrating the method of low-pass filtering initialization, randomized mixed training strategy, and the dip-informed neural network (DINN), we present a self-supervised anti-aliasing algorithm for regular seismic data interpolation, as demonstrated in Fig. 5.7. During the training phase in Fig. 5.7, the low-pass-filtered (LPF-ed) input $\mathbf{z}$ undergoes random decimation (Eq. 5.3) using the new decimation mask $\mathbf{M}_0$ to generate the auxiliary dataset

$$\mathbf{z}' = \mathbf{M}_0\mathbf{z}, \tag{5.10}$$

The randomly decimated dataset $\mathbf{z}'$ and LPF-ed data $\mathbf{z}$ are mixed as a new training dataset which contains both random and regular decimation patterns, and the randomized mix training is accomplished. With the updated training data volume, the combined loss function is developed as:

$$L_\theta(\mathbf{m}, \mathbf{M}_0, \mathbf{L}) = \underbrace{\left\|\mathbf{x} - \sigma_{\mathbf{m}}(f_\theta(\mathcal{F}^{-1}\mathbf{L}\mathcal{F}\mathbf{x}))\right\|}_{\text{LPF-ed loss}} + \underbrace{\left\|\mathbf{M}_0\mathbf{x} - \sigma_{\mathbf{M}_0\mathbf{m}}(f_\theta(\mathbf{M}_0\mathcal{F}^{-1}\mathbf{L}\mathcal{F}\mathbf{x}))\right\|}_{\text{auxiliary loss}}, \tag{5.11}$$

where f_θ represents the DINN parameterized by θ. It is important to note that, while the auxiliary dataset may alter the geographic spacing of the original traces, the authentic wavenumber information is reliably preserved in the original input. This ensures that the network learns this vital information during its training phase. During the inference stage, the auxiliary dataset is no longer utilized. Instead, the only the low-pass-filtered verion of the original data ‡ is used as the input to the network, which enables the restoration of accurate geographic spacing in the reconstructed seismic gathers.

5.3 Analyzing the Randomized Mixed Training Algorithm with DINN

5.3.1 Comparing DINN with the Baseline Methods

The proposed method is evaluated using the same synthetic (SEAM–II unconventional) and field

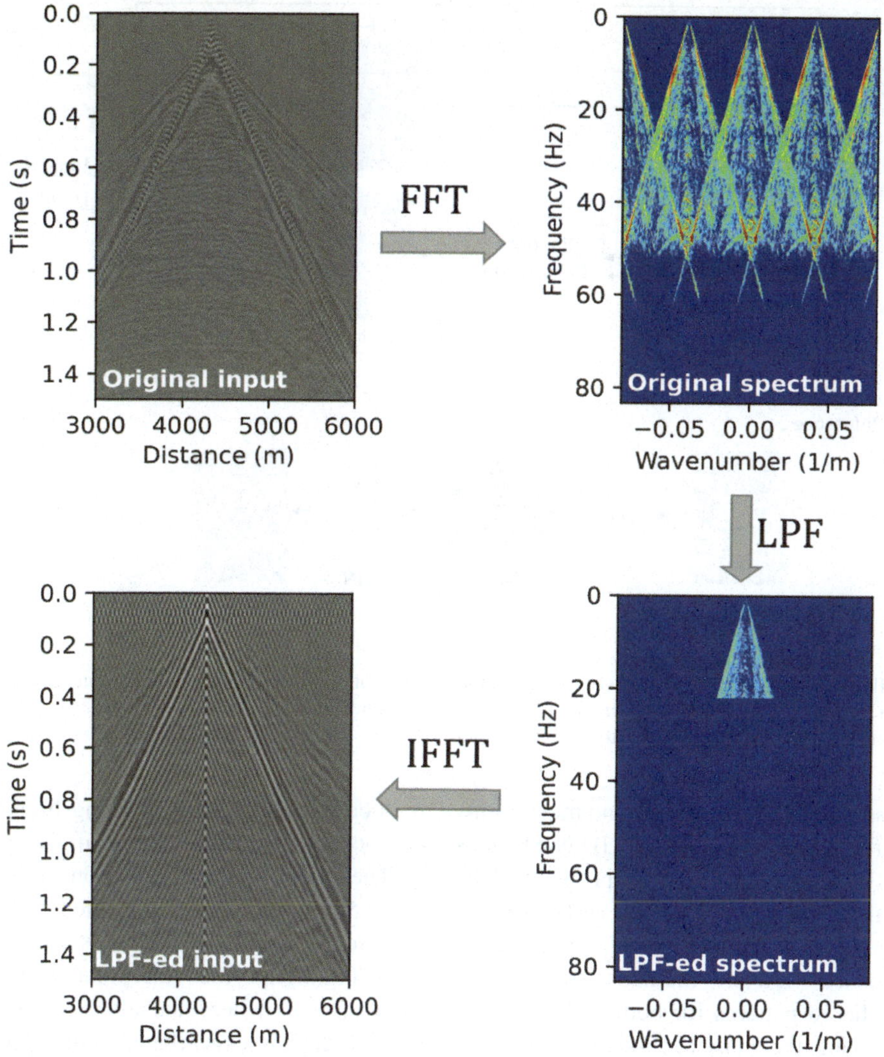

Fig. 5.6 Illustration of low-pass filtering

(Mobile AVO Viking Graben line 12) datasets as discussed in Sect. 4.4.3. As demonstrated in Fig. 5.8, the examples characterized by a 75% regular decimation ratio exhibit a pronounced aliasing issue. As shown in Fig. 5.8d, h, we manually design rectangular f-k masks for the low-pass filtering process to first excessively eliminate the aliased signals from the original input. The LPF-ed results in Fig. 5.8b, f show that both decimated and undecimated traces are initialized with low-frequency components. To create the auxiliary dataset depicted in Fig. 5.7, the original input and LPF-ed data are further randomly decimated by throwing away 20% of the original decimated traces.

To comprehensively assess the effectiveness of our work in addressing the anti-aliasing problem, we compare it with the DSP (Liu et al., 2021) and BTN-SS methods from Sect. 4.4. The metrics for the annotated method on the SEAMII–unconventional dataset are summarized in Table 5.1. The proposed randomized mixed training algorithm with DINN outperforms the other methods in terms of all metrics. Figures 5.9, 5.10, and 5.11 illustrates example predictions, where BTN-SS and DINN both

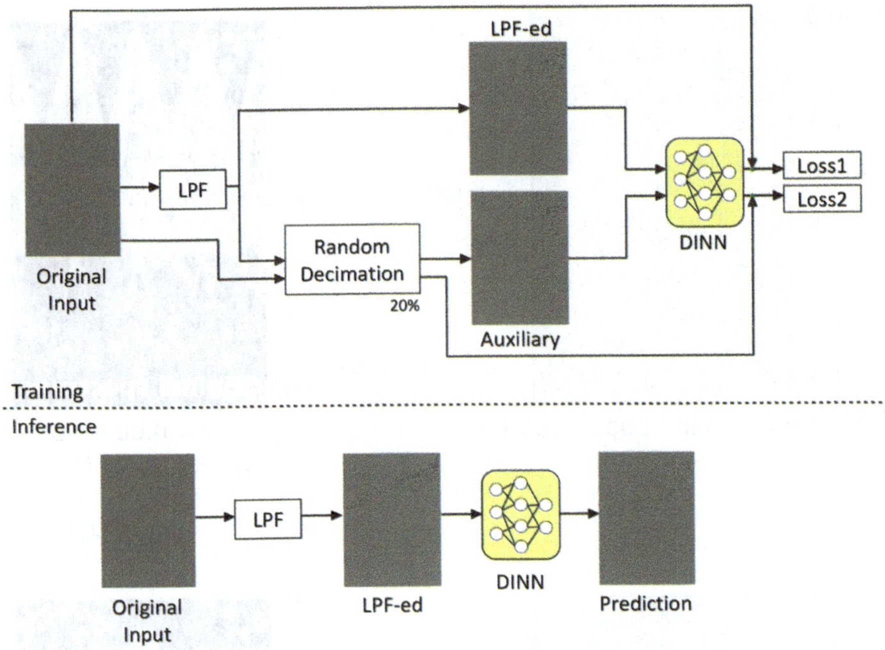

Fig. 5.7 Illustration of the mix training workflow. During training, an auxiliary training data is augmented and combined with the original decimated data though a random decimation process. The auxiliary dataset is discard in the inference phase

yield satisfactory reconstructions and most events are well reconstructed. However, BTN-SS does not fully resolve the aliasing problem. The DSP method only reconstructs early arrivals but fails in the bottom areas. In contrast, the proposed DINN method successfully prevents aliased patterns for both early and late arrivals, giving the most accurate interpolation result.

A more distinct comparison for Figs. 5.9, 5.10, and 5.11 is presented in Fig. 5.12. It reveals that the DSP method, due to its inherent randomness, reconstructs signals correctly and avoids aliasing in this example. However, this is not consistently the case, as the DSP method without explicit dip guidance may struggle with more complex, severely aliased cases, such as the field dataset discussed later. The BTN-SS method fails to eliminate strong aliased components in the high-frequency, low-wavenumber region, mistaking them as non-aliased since it is in the cone-shaped f-k mask. In comparison, our DINN method excels at reconstructing high-frequency components and avoiding aliasing.

The field data evaluation presented in Table 5.2 further underscores the DINN method's superior performance. Figures 5.13, 5.14, and 5.15 highlights the interpolated data of the field dataset, where the aliasing issue for the steeper dips is exacerbated under 75% regular decimation. Here, the near-horizontal patterns in Fig. 5.14c indicate that the DSP method tends to generate low-wavenumber patterns and fails to reconstruct the dealiased components. While BTN-SS method mitigates the aliasing problem to some extent in Fig. 5.14d, some weak aliased energy remains, as indicated by arrows. The DINN method, however, consistently produces the best results in terms of anti-aliasing and accuracy.

The f-k spectra for Fig. 5.13a–e are illustrated in Fig. 5.16. Figure 5.16a demonstrates that the energy is extensively overlapped due to low spatial sampling rate, and only a small portion of it

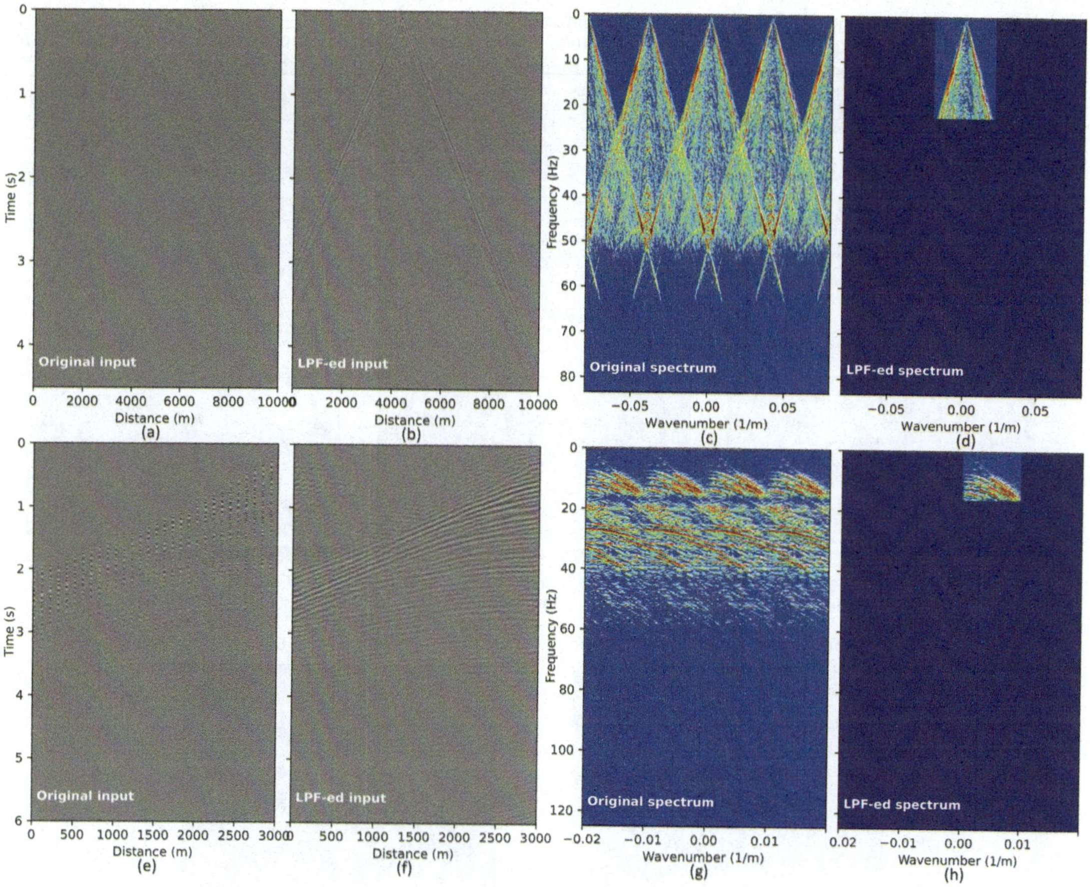

Fig. 5.8 Example of the synthetic and field dataset at different processing stages. (**a**) Regularly sampled SEAM–II unconventional dataset with 75% of the traces missing and filled by 0s; (**b**) The low-pass filtered image of (**a**); (**c**)–(**d**) F-k spectra of the raw input (**a**) and filtered input (**b**), respectively; (**e**)–(**g**) Example of the 75% decimated Mobile AVO Viking Graben line 12 dataset with the same layout as (**a**)–(**d**)

is unaliased and can be used as the initial guess in Fig. 5.8h. The circled part of Fig. 5.16c confirms the inconsistency of the DSP method in anti-aliasing across scenarios. For the BTN-SS result in Fig. 5.16d shows that although the fk-mask suppresses signals outside the cone-shaped area, the signals within that area are not fully recovered. However, the proposed method reconstructs almost all details accurately without aliasing.

Overall, experiments with both synthetic and field data confirm the exceptional effectiveness of our proposed DINN model and randomized mix training algorithm in addressing overfitting and aliasing issues in seismic data interpolation, which consistently outperform existing approaches across various scenarios.

Table 5.1 Performance comparison of the DSP, BTN-SS and DINN method on the 75% sampled SEAM–II unconventional dataset

Methods	Missing ratio	MAE	MSE	SNR
DSP	75%	0.1681	0.1227	1.8588
BTN-SS	75%	0.1045	0.0896	3.2219
DINN	**75%**	**0.0943**	**0.0393**	**6.8088**

The bold values highlight the best results

5.3.2 Effect of the DINN

As elaborated in Sect. 5.2.3, employing low-pass filtering alone can offer an initial estimate of dips, somewhat mitigating the aliasing problem. When combined with randomized mixed training,

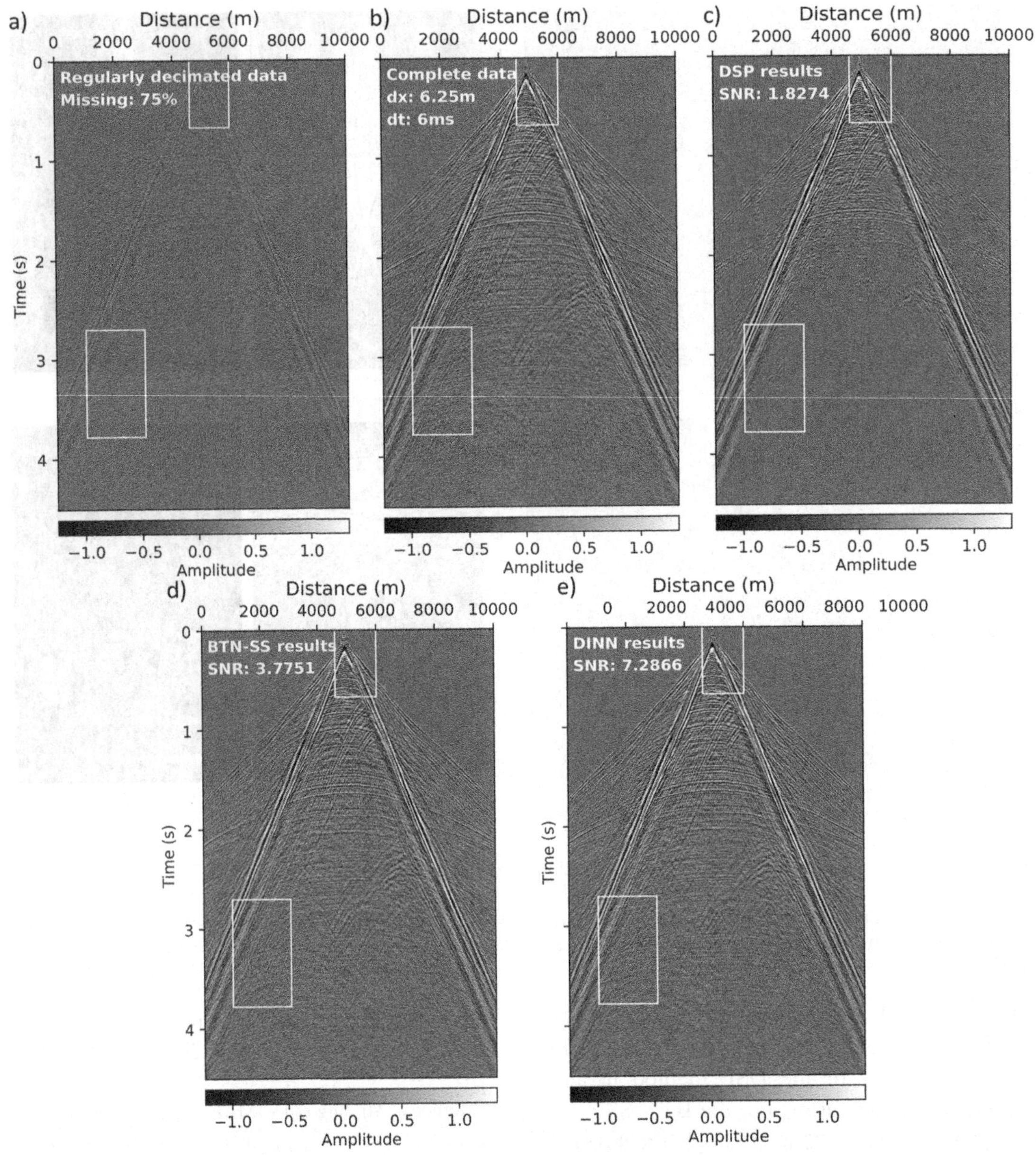

Fig. 5.9 Results of the regularly sampled SEAM-II unconventional dataset with a sampling ratio of 75%. (**a**)–(**b**) Original input with missing traces and complete data; (**c**)–(**e**) Interpolation result by DSP, BTN-SS, and proposed DINN method

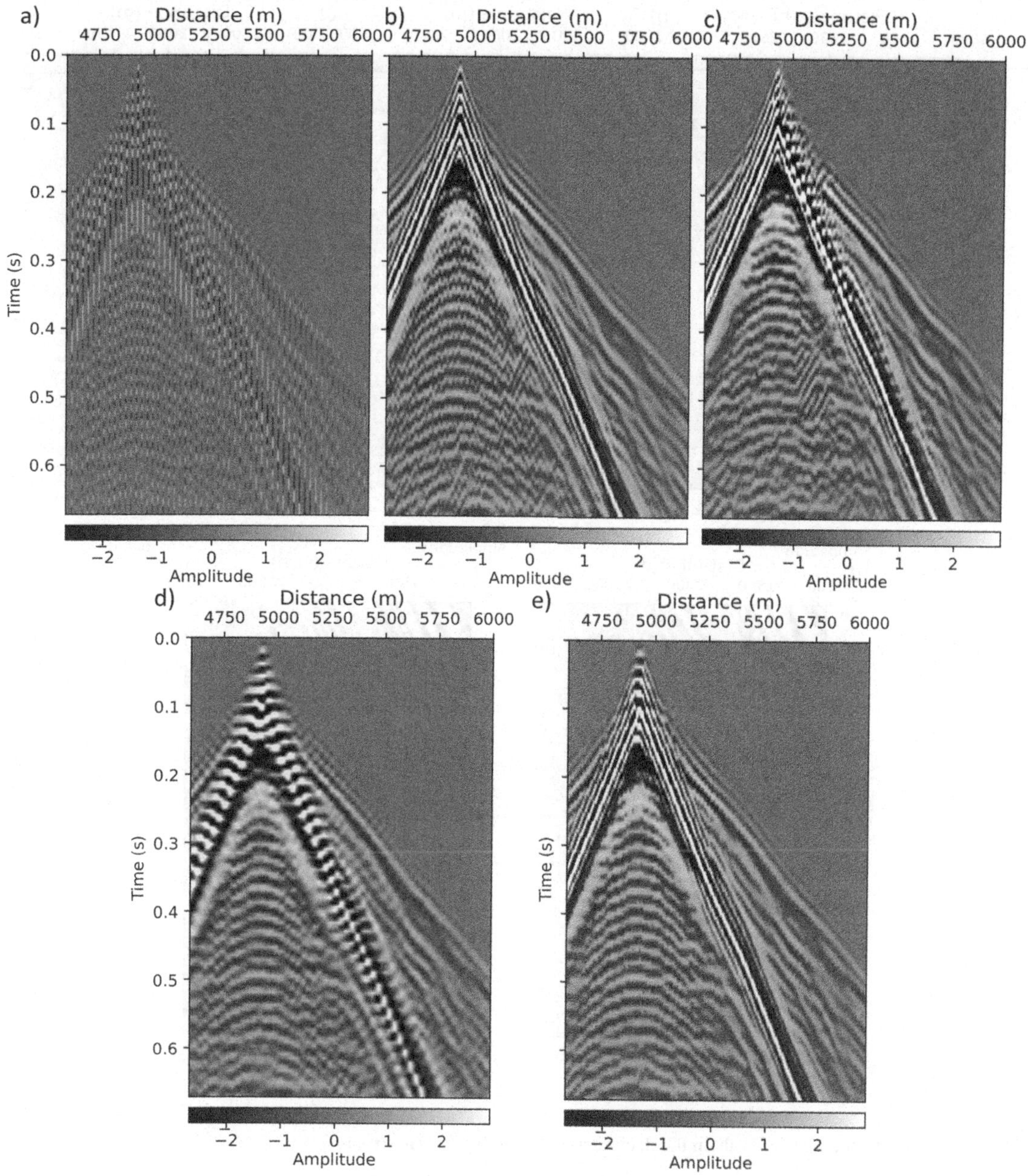

Fig. 5.10 Highlighted early arrivals in the shallow rectangles in Fig. 5.9 (**a**)–(**e**), respectively

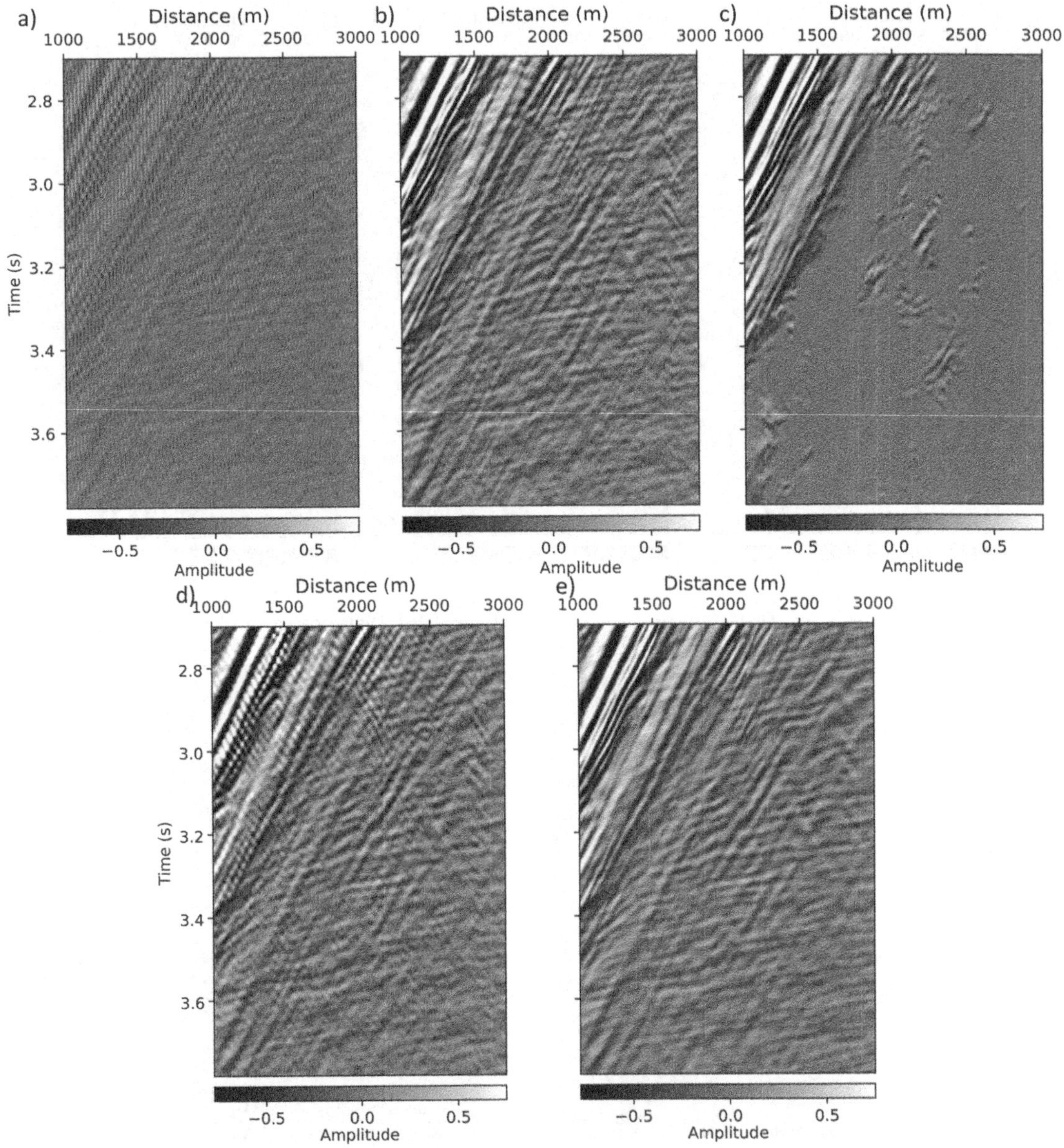

Fig. 5.11 Highlighted late arrivals in the deep rectangles in Fig. 5.9 (**a**)–(**e**), respectively

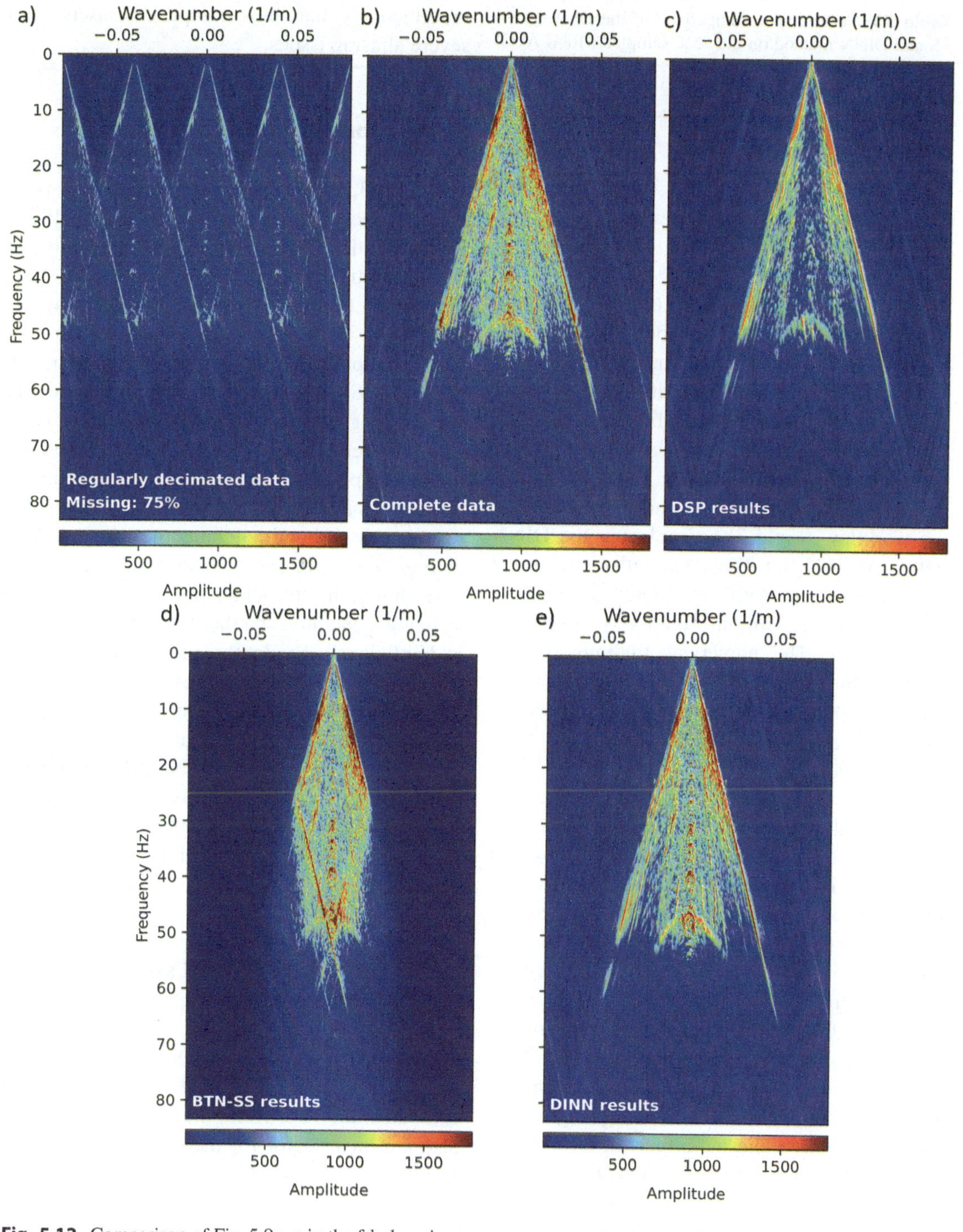

Fig. 5.12 Comparison of Fig. 5.9a–e in the f-k domain

Table 5.2 Performance comparison of the DSP, BTN-SS and DINN method on the 75% sampled Mobil AVO Viking Graben line 12 dataset

Methods	Missing ratio	MAE	MSE	SNR
DSP	75%	0.4155	0.5518	−0.4428
BTN-SS	75%	0.2839	0.2351	3.2750
DINN	**75%**	**0.2517**	**0.2119**	**3.7646**

The bold values highlight the best results

this approach, even without incorporating DINN, shows potential effectiveness for anti-aliasing interpolation. However, its performance might be compromised when dealing with more complex datasets. In this section, we delve into analyzing the impact of the DINN to ascertain its critical role in our proposed workflow.

To facilitate this ablation study, we modified the workflow (Fig. 5.7) by replacing the DINN with a vanilla U-Net. That means all dip-informed deformable convolution (DIDC) layers are avoided and no dip information is explicitly incorporated. The numerical findings of this modification are detailed in Table 5.3 while the qualitative comparisons are presented in Figs. 5.17, 5.18, 5.19, 5.20, 5.21, and 5.22. Observations from Figs. 5.17, 5.18, and 5.19 and their corresponding f-k spectra in Fig. 5.19 reveal striking similarities, suggesting that the integration of dip information does not markedly enhance the results in the SEAM–II unconventional case. This is attributed to the relatively mild aliasing issue in this example, where the initial dip estimation from low-pass filtering provides sufficient dip information, rendering the addition of DINN redundant. Conversely, the scenarios depicted in Figs. 5.20, 5.21, and 5.22 highlight a notable contrast, primarily due to the increased dip angles and more pronounced aliasing issues. In these cases, the workflow without DINN falls short, as evidenced in Figs. 5.20c, 5.21c, and 5.22c, lacking the necessary information and regularization. However, the DINN successfully addresses these challenges, effectively eliminating aliased signals and delivering clean results, as seen in Figs. 5.20d, 5.21d, and 5.22d. This analysis clearly demonstrate the DINN's essential role in effectively handling complex datasets with severe aliasing issues.

5.3.3 Convergence Analysis

In Fig. 5.23, we compare the convergence speed between the baseline methods (DSP and BTN-SS) and the proposed DINN approach on the synthetic dataset.[1] All experiments share the same training configuration and hyper-parameters such as the cuda device, depth of network, optimizer, and learning rate. Although the BTN-SS method achieves the lowest training loss, its high testing loss indicates the severer issue of overfitting caused by aliasing. The DSP method displays a slow convergence in the initial 100 epochs and tends to overfit after 200 epochs, resulting in suboptimal outcomes as indicated in Figs. 5.9, 5.10, and 5.11 and Table 5.1. This trend highlights the challenges associated with utilizing generative methods for both accuracy and efficiency. Evidently, the proposed DINN method present the least overfitting throughout the training process, and achieves the lowest testing loss rapidly compared to the other methods. It shows the DINN and randomized mix training scheme can effectively inform the network with accurate dip information, address the aliasing issue, and efficiently reconstruct the missing traces.

5.3.4 Influence of Dip Estimation Precision

We have shown the importance of integrating dip information into the network. The accuracy of estimated dip map within the DINN workflow is also a critical factor for the network's ability to handle aliasing successfully. This section delves into how inaccuracies in dip map estimations for

[1] The synthetic dataset is adopted in the study as apposed to the field dataset comparing Figs. 5.9 and 5.13, because the DSP method demonstrates certain degree of preventing aliasing on the synthetic dataset and the interpolation result is more acceptable for making it a suitable benchmark.

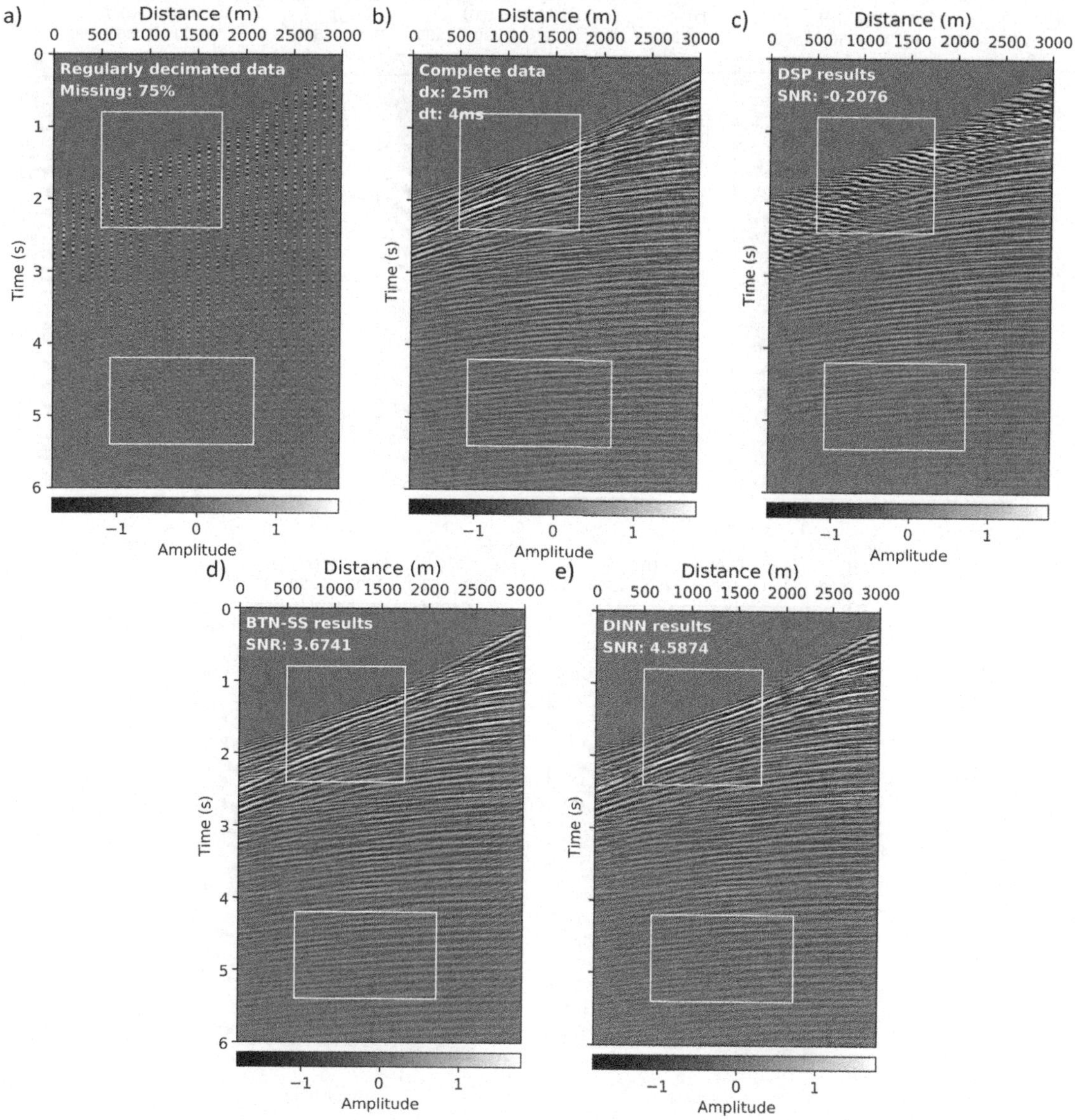

Fig. 5.13 Results of the regularly sampled Mobile AVO Viking Graben line 12 dataset with a sampling ratio of 75%. (**a**)–(**b**) Original input with missing traces and complete data; (**c**)–(**e**) Interpolation result by DSP, BTN-SS, and proposed DINN method

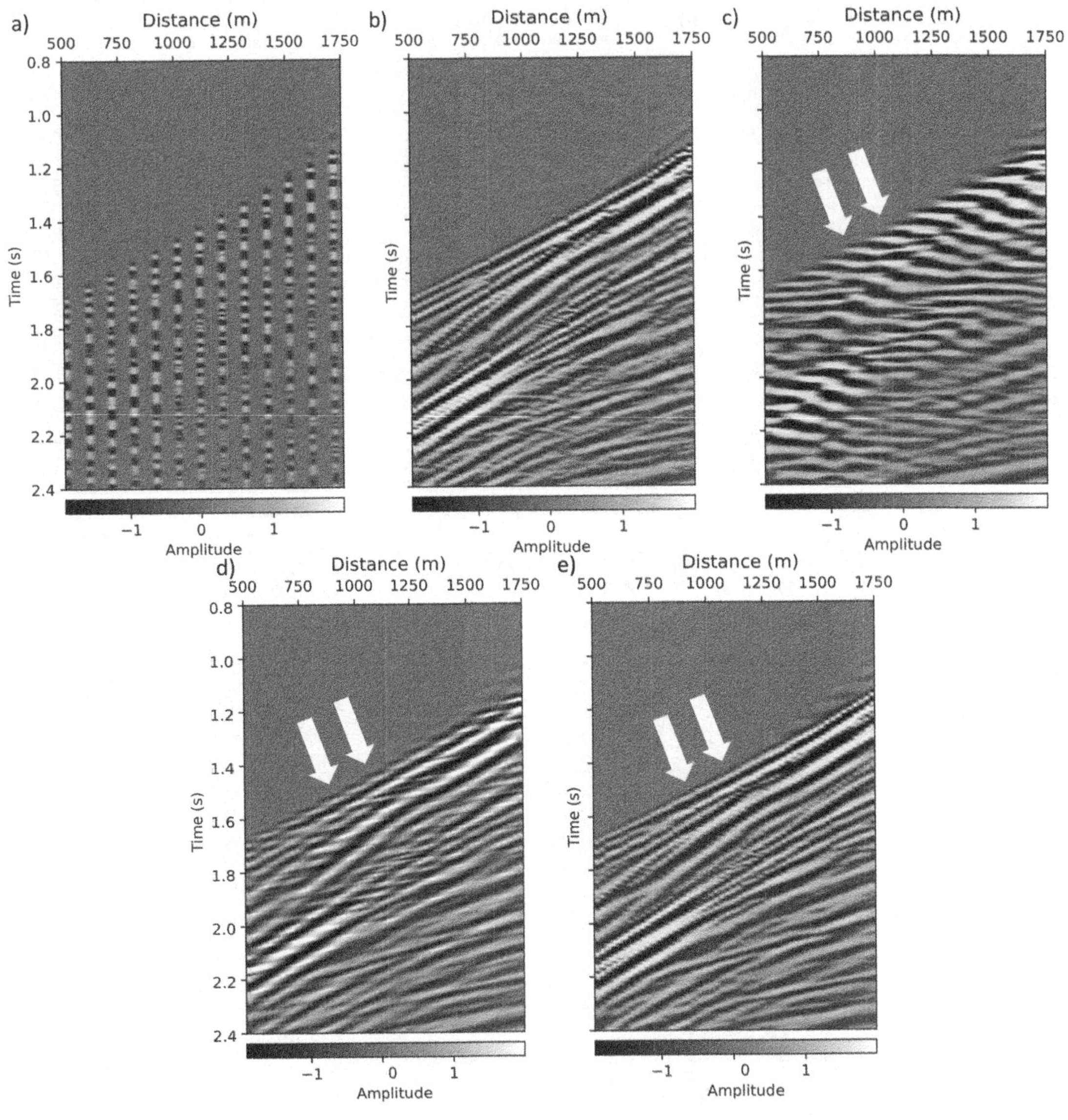

Fig. 5.14 Highlighted early arrivals in the shallow rectangles in Fig. 5.13 (**a**)–(**e**), respectively

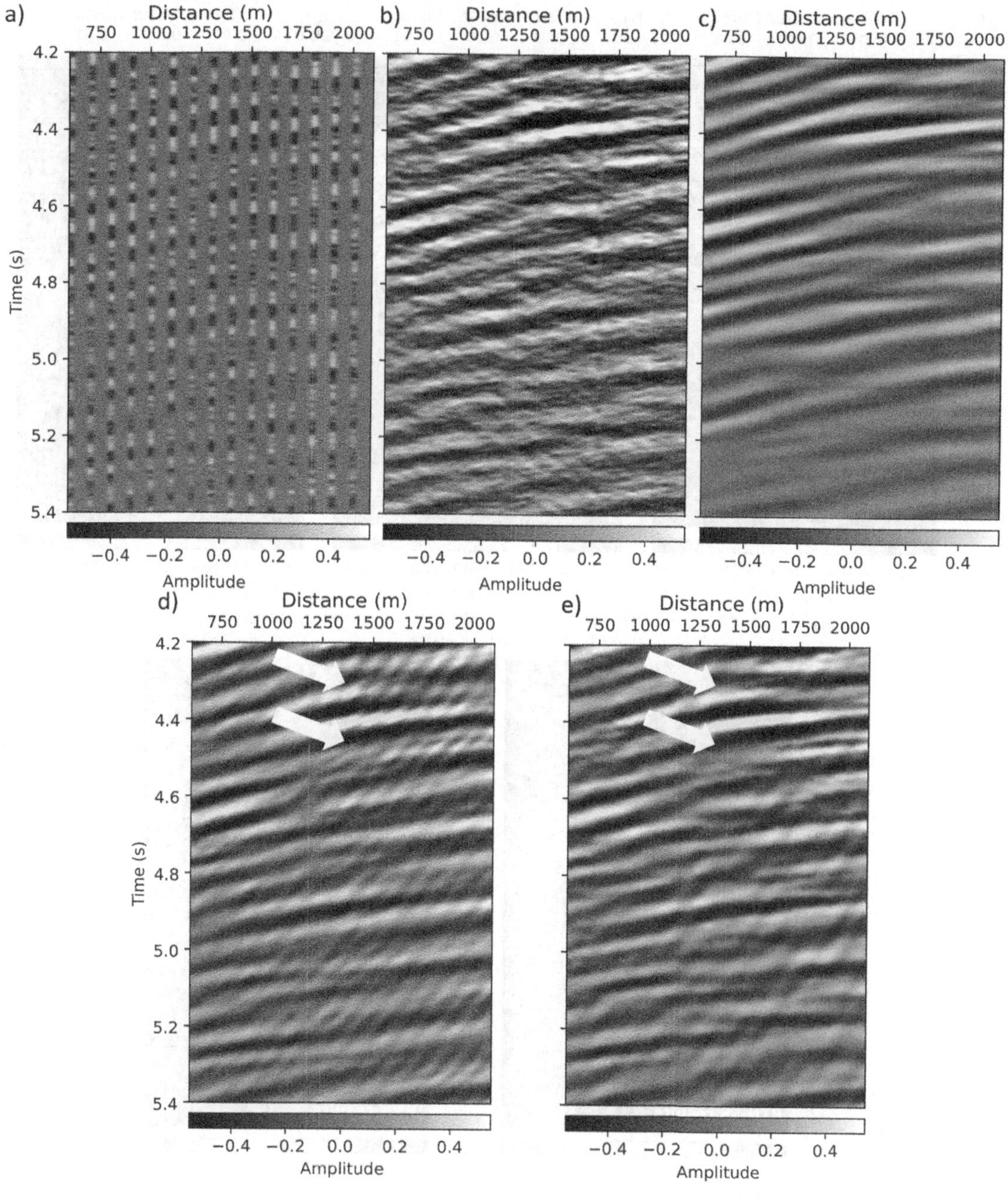

Fig. 5.15 Highlighted late arrivals in the deep rectangles in Fig. 5.13 (**a**)–(**e**), respectively

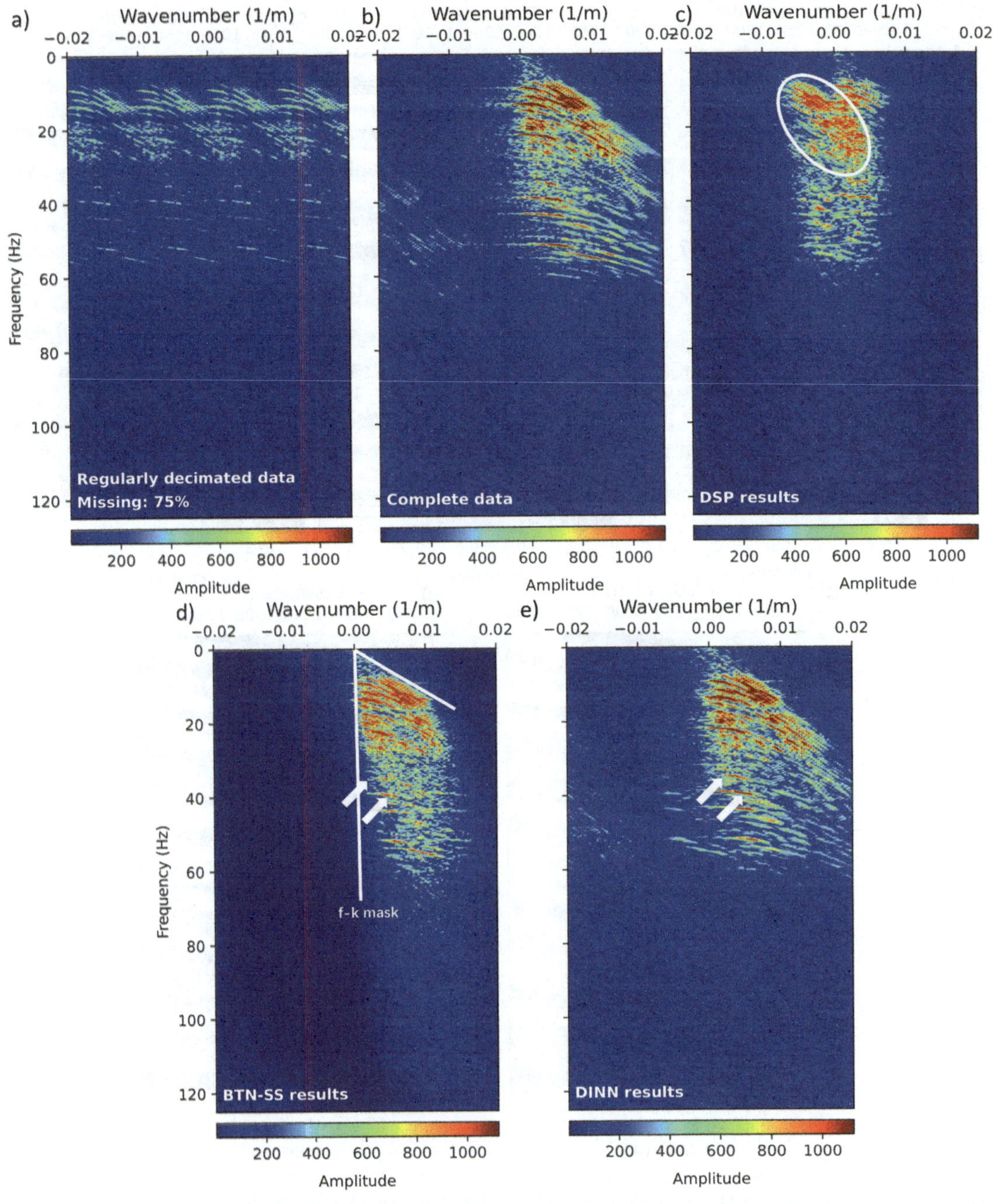

Fig. 5.16 Comparison of Fig. 5.13a–e in the f-k domain

Table 5.3 Comparison between the results of workflow with DINN and without DINN

Datasets	Methods	Missing ratio	MAE	MSE	SNR
SEAMII–unconventional	w/o DINN	75%	0.0921	0.0357	7.2290
	w/ DINN	75%	0.0943	0.0393	6.8088
Mobile AVO Viking Graben line 12	w/o DINN	75%	0.3328	0.4382	0.5857
	w/ DINN	75%	0.2517	0.2119	3.7646

input seismic gathers influence the interpolation outcomes.

Using the field data in this experiment, we generate the dip map from the low-pass filtered (LPF-ed) input, where all gaps are filled such that a continuous dip can be obtained. As shown in Fig. 5.24b, despite its low resolution, the estimated dip map successfully captures the most critical dips in the input data highlighted by arrows, where aliasing is predominantly likely due to the steepest slopes. Subsequently, we apply a scaling factor λ decreasing from 1.0 to 0.8 across the map (as illustrated in Fig. 5.24c–e), which progressively introduces errors and diminishes the influence of the dip map.

The corresponding interpolation results are displayed in Figs. 5.25, 5.26, and 5.27 and the loss variations is depicted in Fig. 5.28. It's important to note that, although the dip angles at early arrivals may not appear steep visually due to visualization stretching, they can reach over 70 degrees that cause a hgih risk of aliasing. A reduction in training loss coupled with an increase in testing loss, as shown in Fig. 5.28, suggests an overfitting problem linked to aliasing deterioration with reduced dip map accuracy. Figure 5.25a–e demonstrates an escalation in aliasing for steep early arrival events. However, for more moderate slopes, indicated by the circular area, the decline is less pronounced. In the case of late arrivals in Fig. 5.26a–e, no significant degradation is observed. This implies that dip map inaccuracies predominantly affect the handling of steep, critical events, while their impact on the broader reconstruction of seismic data remains minimal.

It is worth noting that the adoption of single-channel pixel-wise dip maps generally results in an emphasis on the more prominent dips within seismic data. However, it is worthy noting that the proposed approach has shown proficiency in dealing with conflicting dips in scenarios with neutral dip overlapping. A clear example is observed in Fig. 5.11e, where the network effectively distinguishes and retains two significant dips spread across a large section of the seismic gather. On the other hand, in scenarios of higher complexity, like the one shown in Fig. 5.10e, characterized by finer and more densely interfering events, tends to focus on reconstructing the dominant, lower-wavenumber dips. To tackle these intricate cases, the employment of a multi-channel dip map offering higher resolution would be beneficial. This would equip the network with a broader range of dip information, enhancing its ability to accurately decode and reconstruct the detailed nuances present in seismic data.

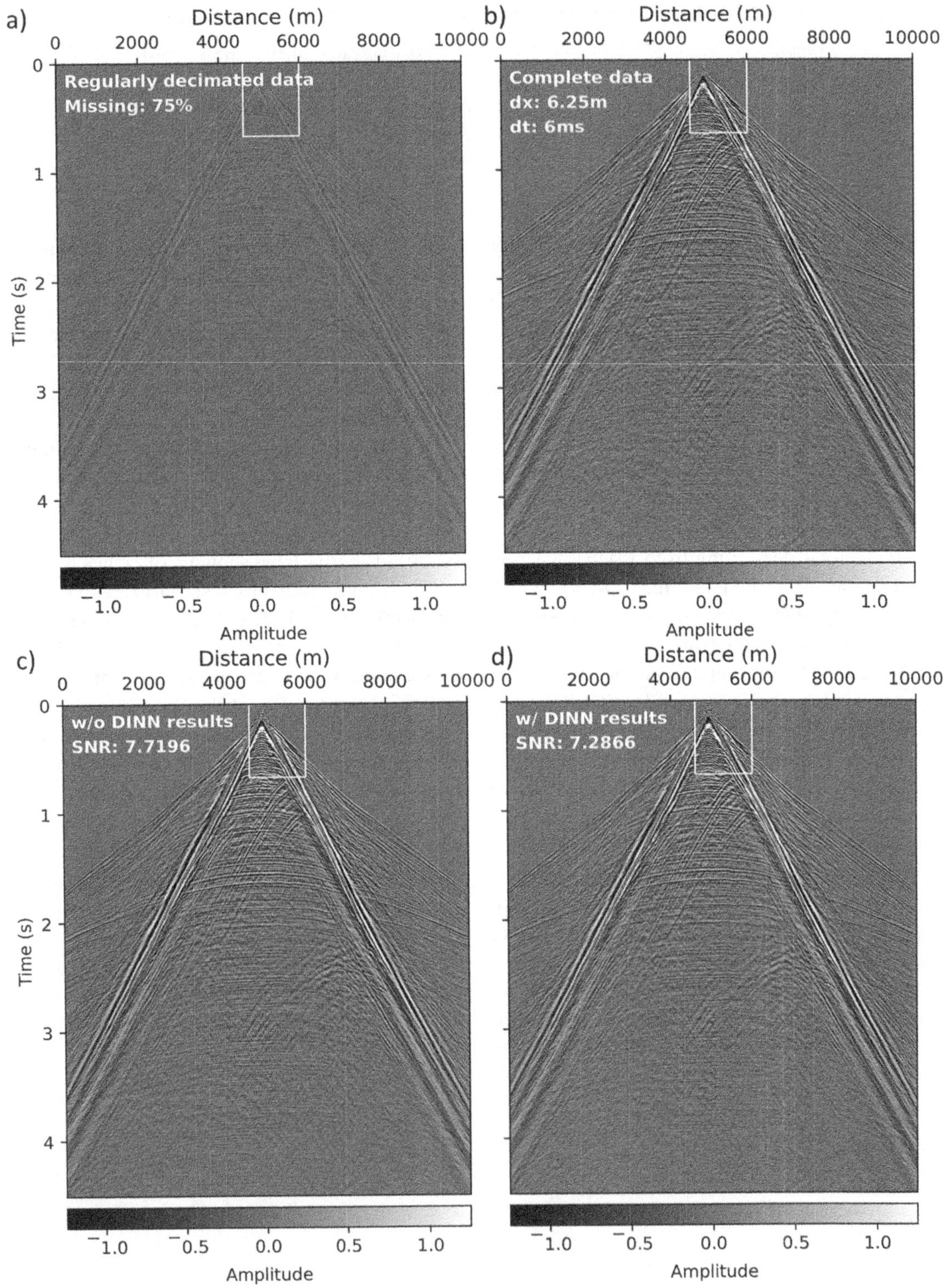

Fig. 5.17 Ablation study of the proposed DINN on the synthetic dataset with a sampling ratio of 75%. (**a**)–(**d**) Decimated seismic input, Complete data, reconstructed data using normal U-Net, and reconstructed data using DINN

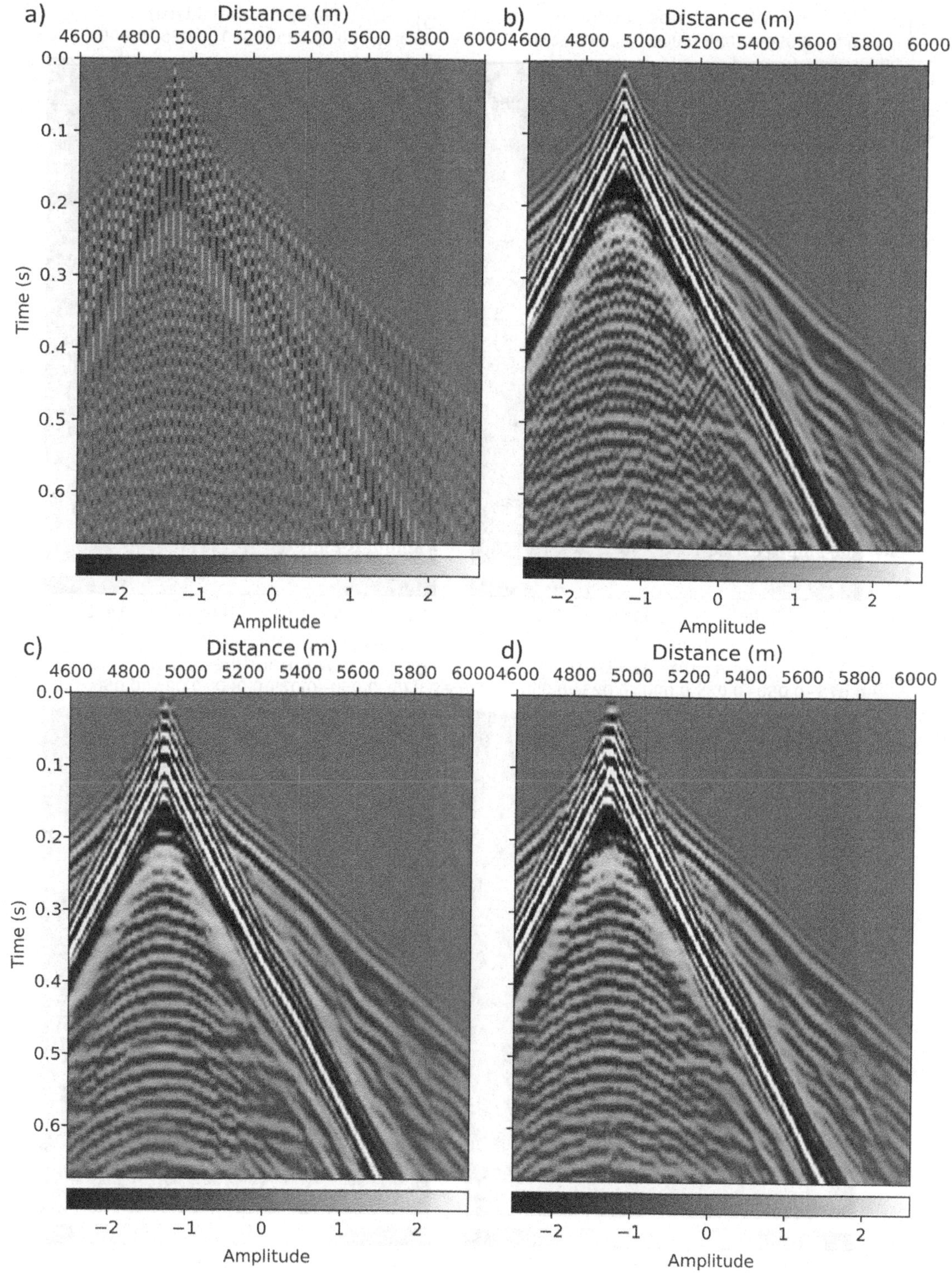

Fig. 5.18 Highlighted area in the rectangles in Fig. 5.17, respectively

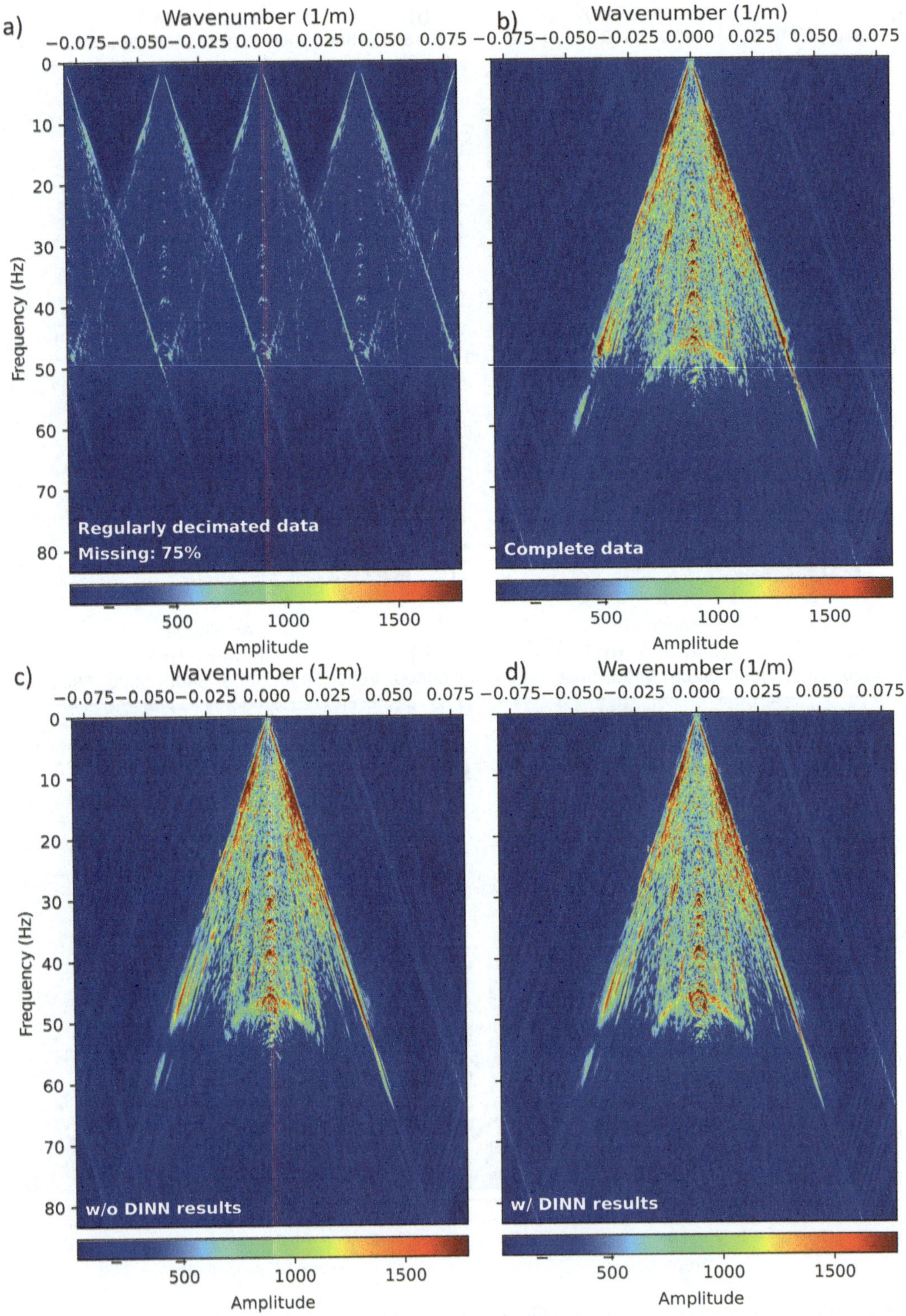

Fig. 5.19 F-k spectra of Fig. 5.17 (**a**)–(**d**), respectively

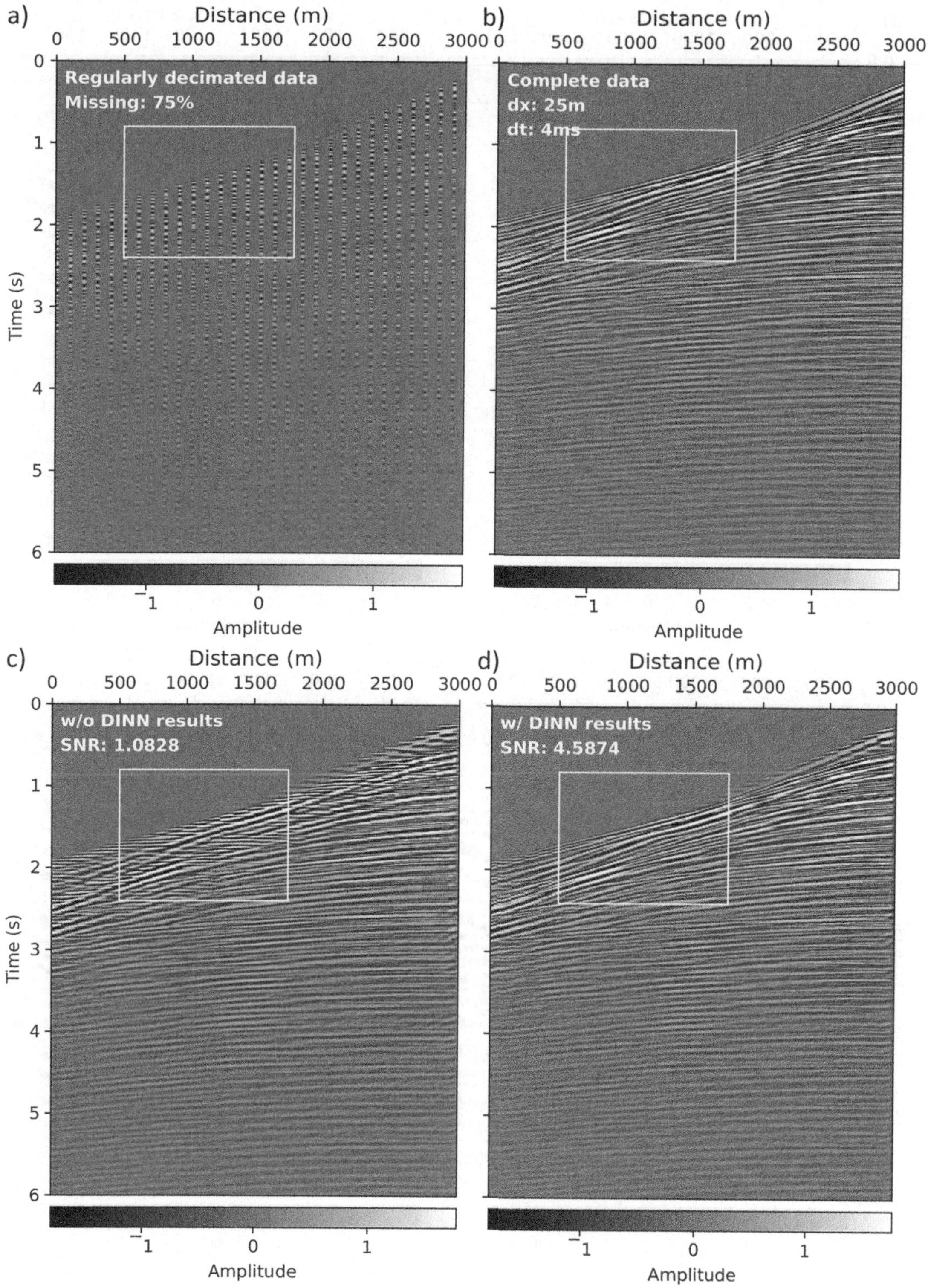

Fig. 5.20 Ablation study of the proposed DINN on the field dataset with a sampling ratio of 75%. (**a**)–(**d**) Decimated seismic input, Complete data, reconstructed data using normal U-Net, and reconstructed data using DINN

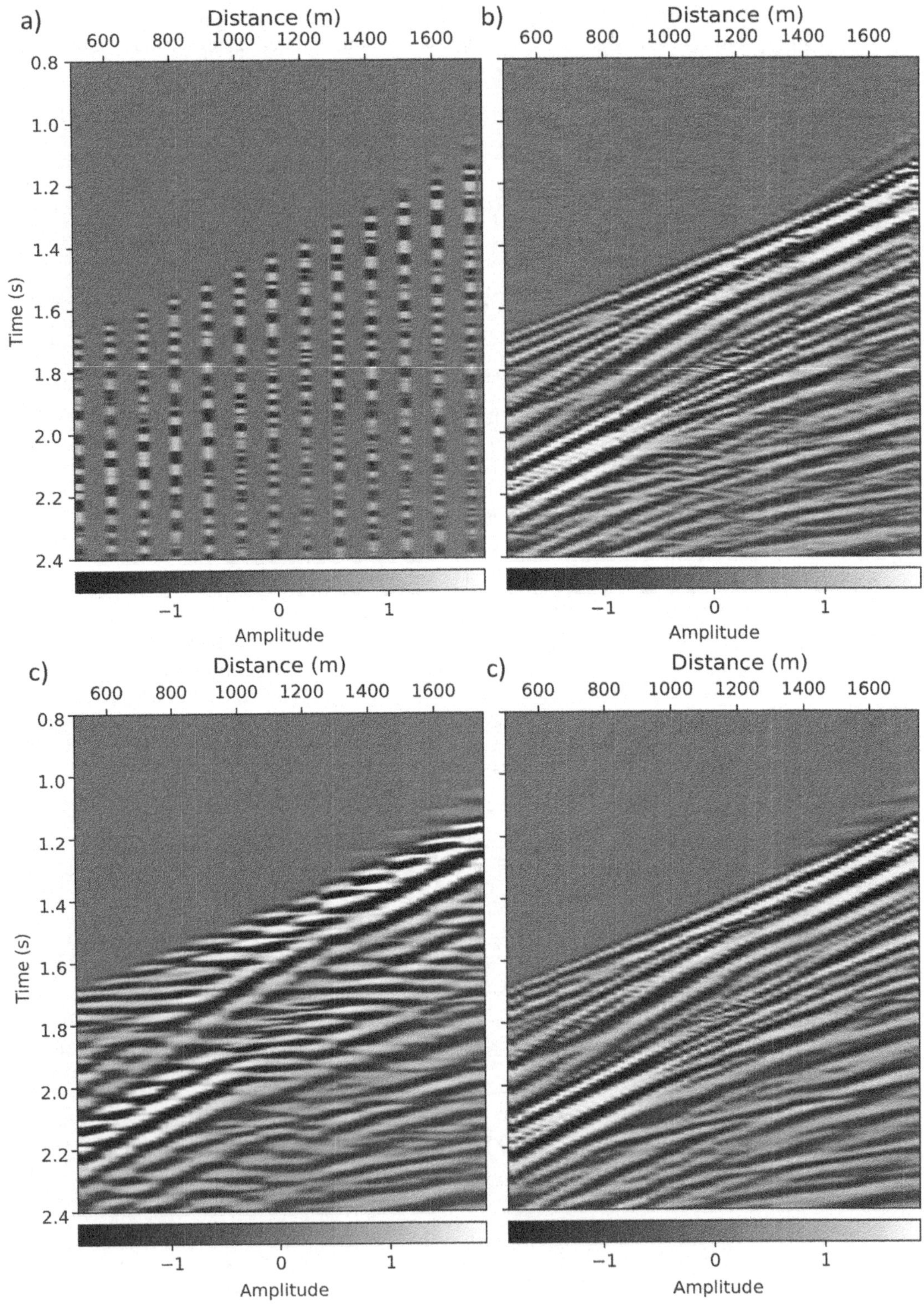

Fig. 5.21 Highlighted area in the rectangles in Fig. 5.20, respectively

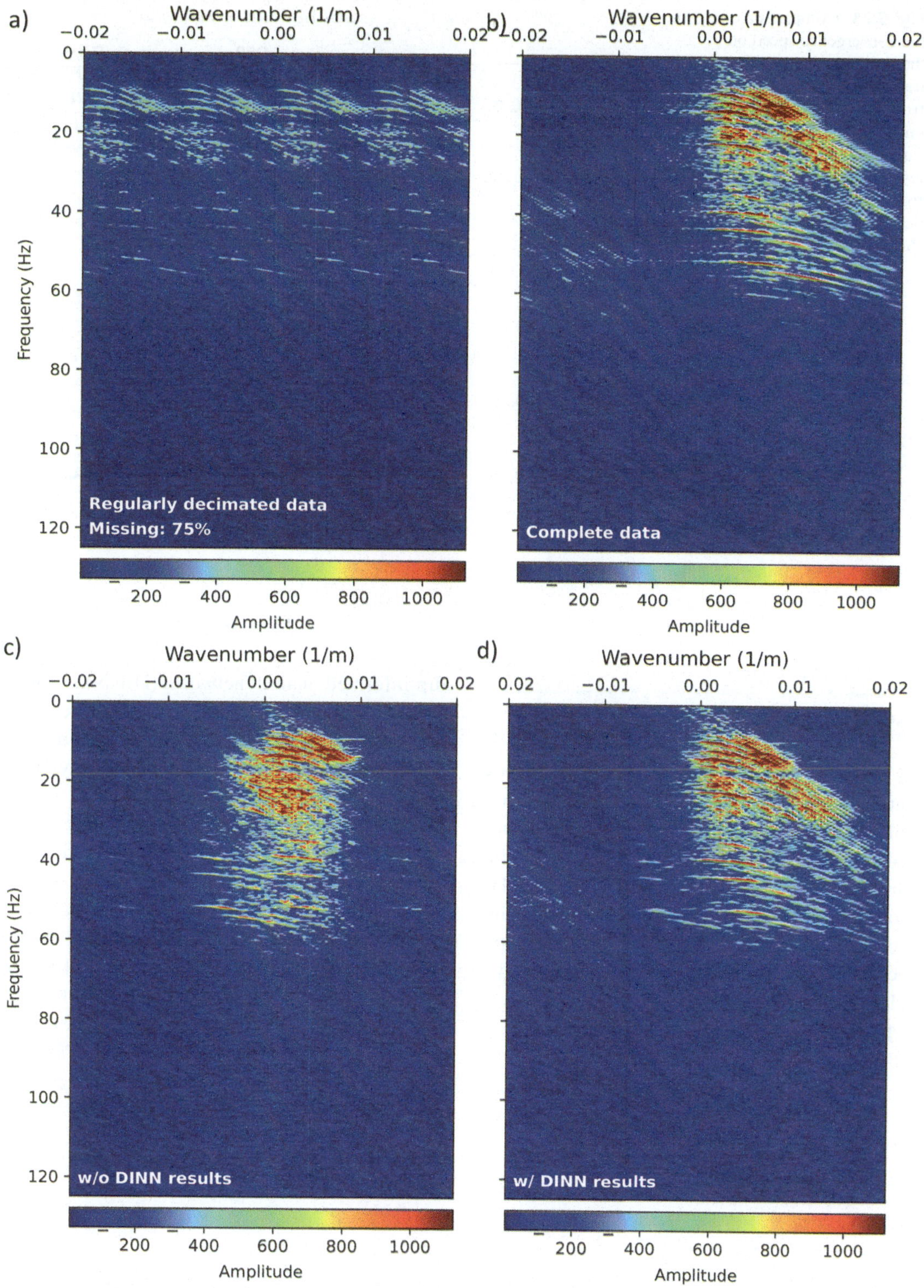

Fig. 5.22 F-k spectra of Fig. 5.20 (**a**)–(**d**), respectively

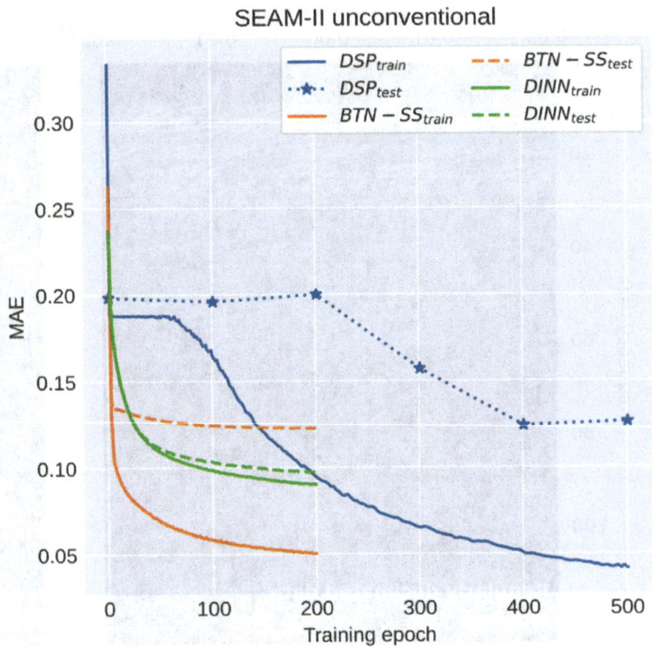

Fig. 5.23 Comparison of the convergence speed on the synthetic dataset. The training curve is obtained from the reconstruction of undecimated traces by Eq. (5.1), whereas the testing curve is obtained from the reconstruction of decimated traces by Eq. (5.2)

5.4 Summary

In this chapter, we introduced an innovative anti-aliasing self-supervised learning algorithm tailored for regularly sampled seismic data interpolation. Our approach begins with the creation of a randomized mixed training workflow. This workflow is designed to infuse randomness into the decimation pattern, effectively mitigating the overfitting issue commonly encountered in self-supervised interpolation tasks involving regular data. Subsequently, we introduce the dip-informed neural network (DINN), a novel solution specifically crafted to address the aliasing challenge. The efficacy of the proposed approach is rigorously validated by both synthetic and field datasets, characterized by a 75% decimation ratio and significant aliasing. Comparative analyses with existing methodologies underscore the enhanced performance of our approach. Additionally, the conducted ablation studies sheds light on the critical role of dip information within the DINN, further affirming the method's reliability and effectiveness in seismic data interpolation.

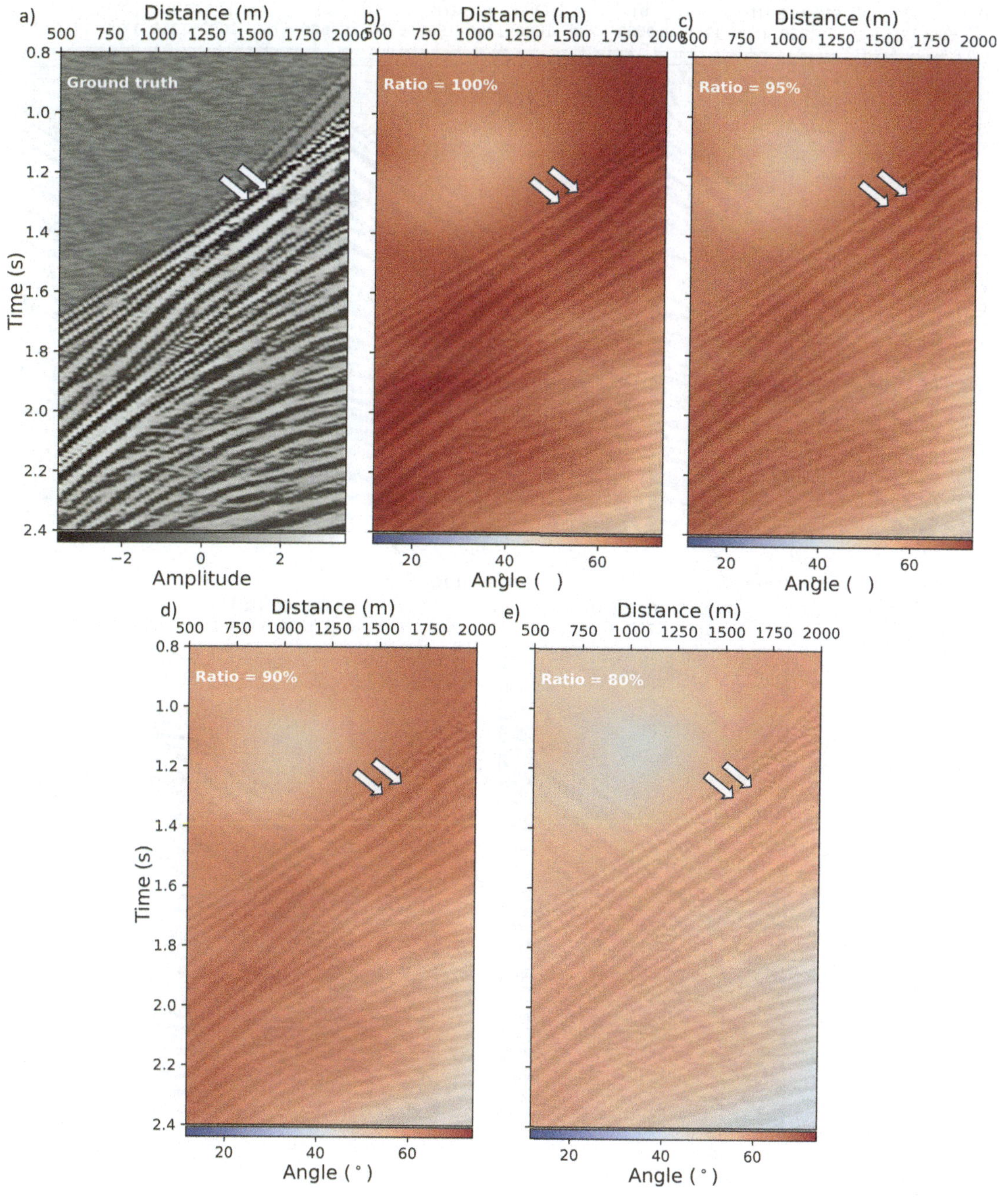

Fig. 5.24 Illustration of the effective angles when applying varied scaling factor λ to the estimated angle map by PWD. (**a**) A field data exmple; (**b**) The original angle map from PWD ($\lambda = 1.0$); (**c**)–(**d**) Angle maps when applying λ decreasing from 0.95 to 0.8

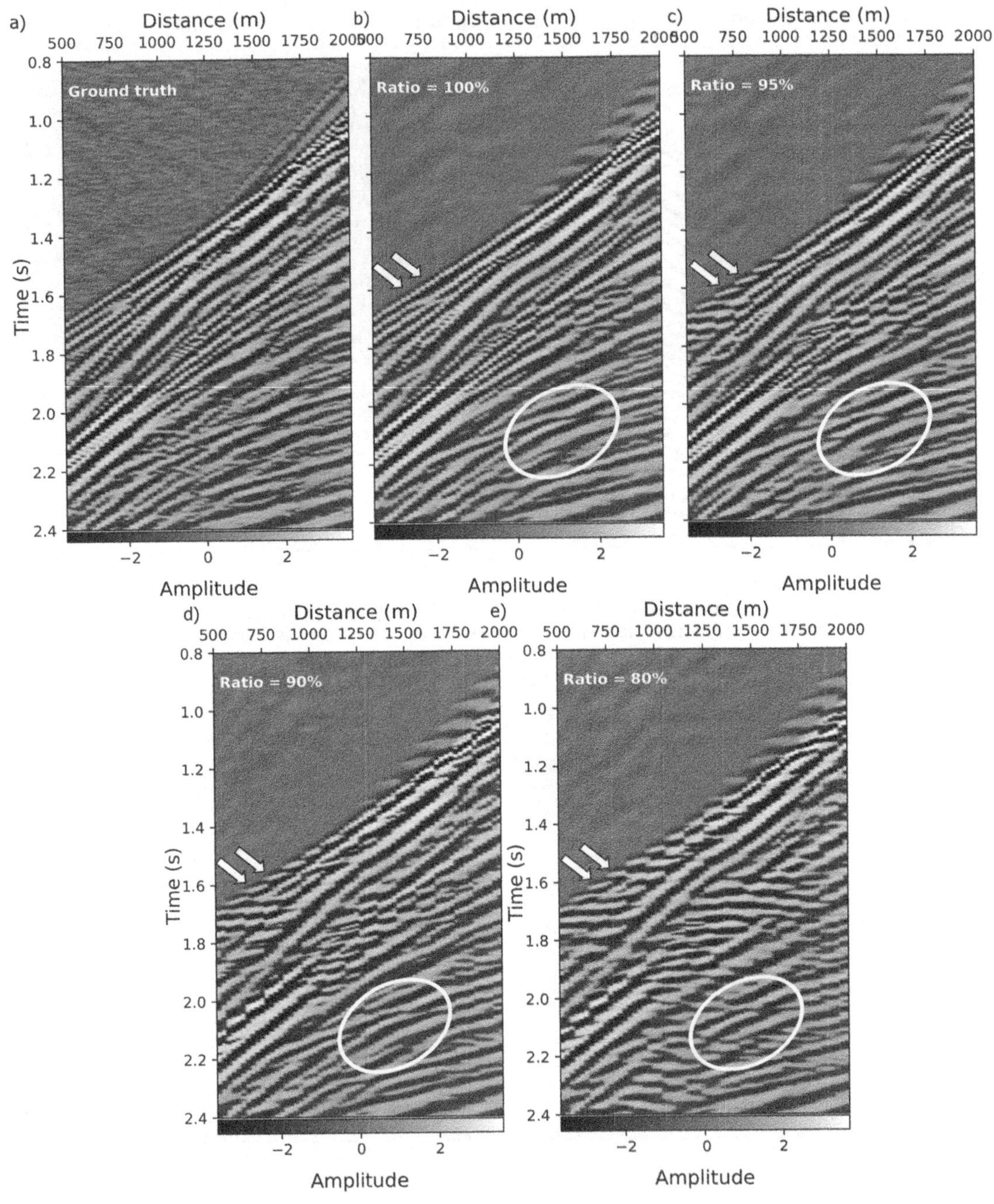

Fig. 5.25 Results of the field dataset with a sampling ratio of 75% when gradually diminishing the impact of the angle maps. (**a**)–(**e**) The complete seismic label and interpolation result when the estimated angle map is scaled by a $\lambda = 1.0, 0.95, 0.9, 0.8$, respectively

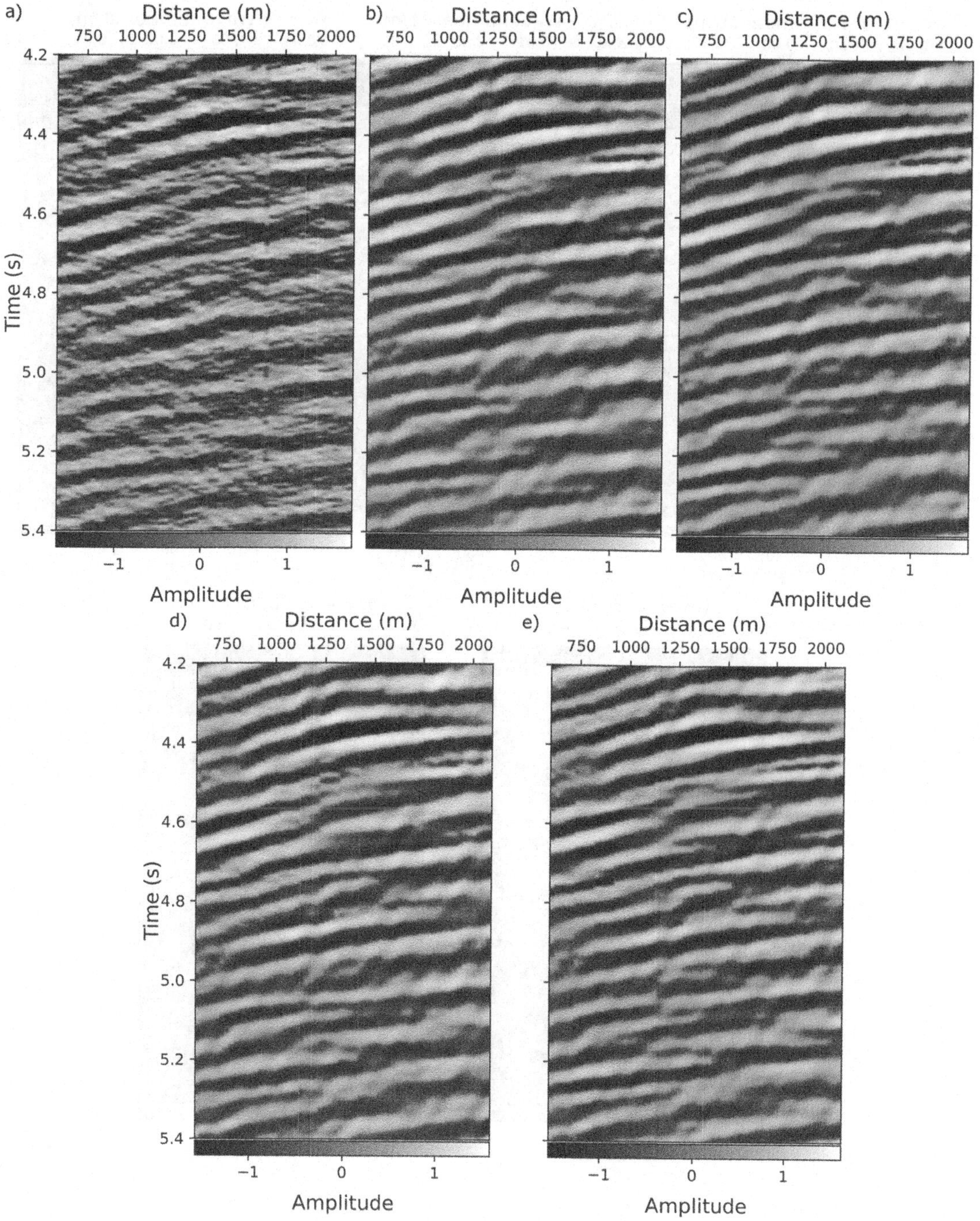

Fig. 5.26 Highlighted areas in Fig. 5.25 (**a**)–(**e**), respectively

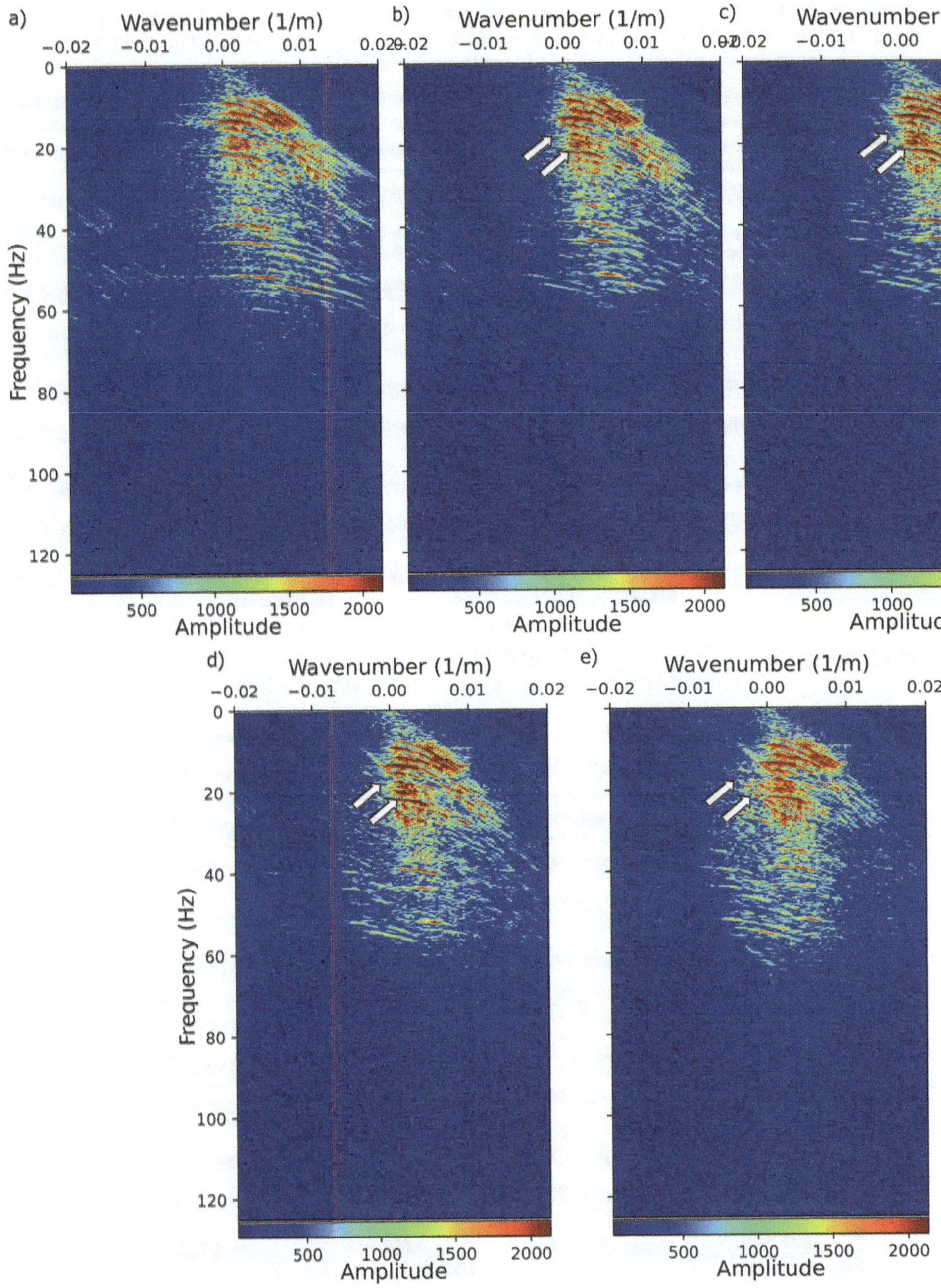

Fig. 5.27 F-k spectra for Fig. 5.25 (**a**)–(**e**), respectively

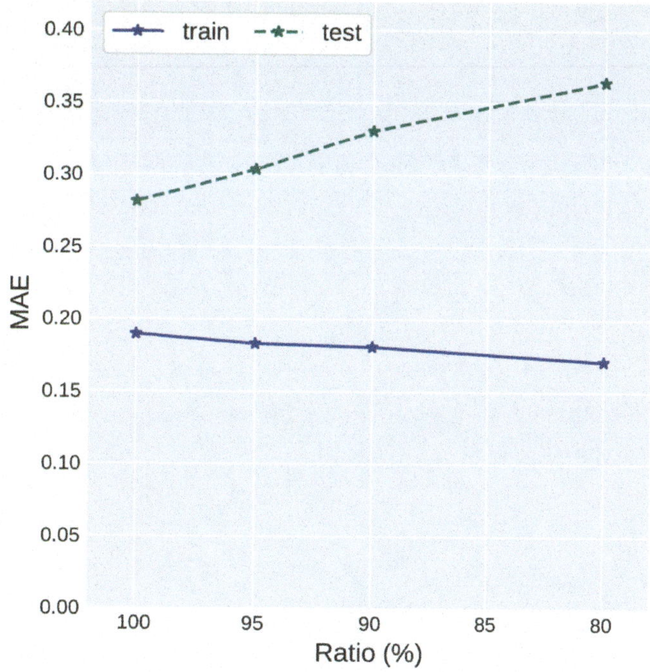

Fig. 5.28 Variation of performance when gradually diminishing the impact of the dip map. The scaling factor λ is decreased from 1.0 to 0.8 along the horizontal axis. The metrics are measured in after the networks are sufficiently trained

References

Abma, R., Kabir, N.: 3D interpolation of irregular data with a POCS algorithm. Geophysics **71**(6), E91–E97 (2006)

Curry, W.: Interpolation with fourier-radial adaptive thresholding. Geophysics **75**(6), WB95–WB102 (2010)

Dai, J., Qi, H., Xiong, Y., Li, Y., Zhang, G., Hu, H., Wei, Y.: Deformable convolutional networks. In: Proceedings of the IEEE International Conference on Computer Vision, pp. 764–773 (2017)

Fang, W., Fu, L., Liu, S., Li, H.: Dealiased seismic data interpolation using a deep-learning-based prediction-error filter. Geophysics **86**(4), V317–V328 (2021)

Fomel, S.: Applications of plane-wave destruction filters. Geophysics **67**(6), 1946–1960 (2002)

Fomel, S.: Towards the seislet transform. In: 2006 SEG Annual Meeting. OnePetro, Richardson (2006)

Fomel, S., Liu, Y.: Seislet transform and seislet frame. Geophysics **75**(3), V25–V38 (2010)

Gan, S., Wang, S., Chen, Y., Zhang, Y., Jin, Z.: Dealiased seismic data interpolation using seislet transform with low-frequency constraint. IEEE Geosci. Remote Sens. Lett. **12**(10), 2150–2154 (2015)

Gülünay, N.: Seismic trace interpolation in the fourier transform domain. Geophysics **68**(1), 355–369 (2003)

He, K., Zhang, X., Ren, S., Sun, J.: Deep residual learning for image recognition. In: Proceedings of the IEEE Conference on Computer Vision and Pattern Recognition, pp. 770–778 (2016)

Liu, Q., Fu, L., Zhang, M.: Deep-seismic-prior-based reconstruction of seismic data using convolutional neural networks. Geophysics **86**(2), V131–V142 (2021)

Naghizadeh, M.: Seismic data interpolation and denoising in the frequency-wavenumber domain. Geophysics **77**(2), V71–V80 (2012)

Naghizadeh, M., Sacchi, M.D.: Beyond alias hierarchical scale curvelet interpolation of regularly and irregularly sampled seismic data. Geophysics **75**(6), WB189–WB202 (2010)

Picetti, F., Lipari, V., Bestagini, P., Tubaro, S.: Anti-aliasing add-on for deep prior seismic data interpolation. In: 2021 IEEE International Conference on Image Processing, pp. 1979–1983. IEEE, Piscataway (2021)

Ronneberger, O., Fischer, P., Brox, T.: U-net: convolutional networks for biomedical image segmentation. In: International Conference on Medical Image Computing and Computer-Assisted Intervention, pp. 234–241. Springer, Berlin (2015)

Wang, B., Zhang, N., Lu, W., Wang, J.: Deep-learning-based seismic data interpolation: a preliminary result. Geophysics **84**(1), V11–V20 (2019)

Zhang, H., Zhang, H., Zhang, J., Hao, Y., Wang, B.: An anti-aliasing POCS interpolation method for regularly undersampled seismic data using curvelet transform. J. Appl. Geophys. **172**, 103894 (2020)

6 Deep Learning for Seismic Data Compression

6.1 Introduction to Seismic Data Compression

The seismic industry has experienced a significant increase in data volume in recent years, primarily attributed to advances in digital technology and seismic data acquisition methods. This explosive growth poses considerable challenges, notably in data storage, transmission, and processing. Addressing these challenges accelerates the development of an efficient data compression algorithm tailored to leverage the inherent sparsity and redundancy in seismic data. This approach facilitates a more compact representation of the data by effectively reducing its dimensions. While certain compression techniques aim for lossless results, the specific characteristics of seismic data—marked by redundancy and noise—make lossy compression algorithms a more suitable choice. These algorithms, by tolerating a certain level of data loss, manage to preserve the essential quality and interpretability of seismic data without significant compromise.

Seismic data compression has been longstanding subject of discussion, but its importance has grown substantially in light of the exponential increase in seismic data volume. In this context, the transform-based compression algorithms play a pivotal role. As illustrated in Fig. 6.1, these algorithms typically begin with a record $\mathbf{x}$ and encode it into a new vector $\mathbf{y}$. This encoding occurs in a domain where the signals exhibit sparse representation and minimal redundancy. The compressed form of $\mathbf{y}$—sparse, coded, and dimensionally reduced—is then stored or transmitted. A decompressor later retrieves this representation, decoding and converting $\mathbf{y}$ back into its original form, now denoted as $\hat{\mathbf{x}}$. Depending on specific needs and applications, the encoding and decoding stages can be adjusted or omitted.

Various conventional transformation techniques have been applied in seismic data compression. These include the Discrete Fourier Transform (DFT), the Discrete Cosine Transform (DCT) (Spanias et al., 1991; Wang and Wu, 2000), the Walsh-Hadamard Transform (WHT) (Wood, 1974), the Wavelet Transform (WT)(Reiter and Heller, 1994; Duval and Nguyen, 1999), and the Seislet Transform (ST) (Fomel, 2006). Each of these methods offers distinct approaches to effectively manage the complexities involved in compressing seismic data.

Recently, machine learning has shown great promise in natural image compression (Ballé et al., 2016; Rippel and Bourdev, 2017; Mentzer et al., 2018; Ballé et al., 2018). These contemporary methods stand in contrast to traditional image or video compression algorithms used by JPEG, JPEG2000, and HEVC. The key distinction lies in the replacement of conventional transformation processes with fully parameterized neural networks, which can be optimized for specific

S. Wang et al., *Deep Learning for Seismic Data Enhancement and Representation*, Advances in Oil and Gas Exploration & Production, https://doi.org/10.1007/978-3-031-75745-7_6

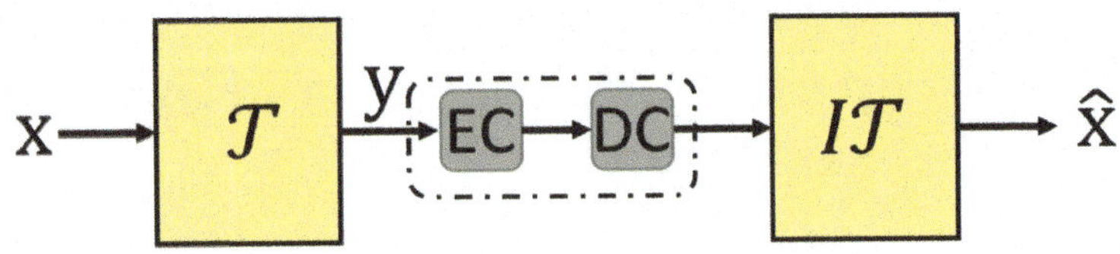

Fig. 6.1 Workflow of transform-based image compression. An image **x** is transformed into a latent space by $\mathcal{T}$, where the representation is coded into bit stream for storage and transmission. In the decompression stage, the binary codes are decoded into the latent vector and the inverse transformation $I\mathcal{T}$ is applied to restore the image $\hat{\mathbf{x}}$

target datasets. The primary advantage of these data-driven approaches is their ability to learn and adapt to the unique structures of different datasets. This capability enables the creation of customized compressors that are specifically tailored to the characteristics of each dataset. As a result, machine learning-based methods offer a more efficient and effective solution for image compression, particularly when dealing with diverse and complex data structures.

There are multiple types of seismic data to be compressed at different processing stages, including pre-migration, post-migration, and intermediate data during computations. For different stages, the data complexity, coherency, and noise level can be very different. In this chapter, we focus on compressing the pre-stack, pre-migration, and post-migration data in a 2D manner. Huang (1999) adopted a neural network to find the principal components of seismograms. Nuha et al. (2018) used a shallow neural network with one hidden layer to compress the seismic traces. Then they proposed a deeper autoencoder-based model to compress seismic data, trained with an extreme learning machine (ELM). Helal et al. (2021) proposed a convolutional autoencoder (CAE) based method for seismic trace compression. Two deep CNNs are constructed to achieve varied compression ratios. Most current literature compresses the seismic data in a 1D manner and there are few methods to leverage the spatial coherency of seismic data and formulate it as higher-dimensional compression. Schiavon et al. (2019) used a 2D autoencoder to compress post-migration seismic data and achieve promising results, and they extended it to the 3D version (Schiavon et al., 2020).

Seismic data compression varies across stages, posing different challenges in complexity, coherency, and noise. In this chapter, our focus is on compressing pre-stack, pre-migration, and post-migration data, specifically using a 2D approach. This focus allows for a more nuanced exploration of the data's spatial characteristics, which can significantly enhance the efficiency and efficacy of compression techniques in seismic data processing.

6.2 Variational Autoencoder with Hyperprior for Seismic Data Compression

6.2.1 Rate Distortion Trade-Off

A high compression ratio, which reflects the ratio of data occupancy before and after compression, is typically desirable in seismic data compression due to the significant internal correlations among signals and the prevalent noise. Nonetheless, this goal needs to be carefully balanced with the need to maintain decompressed data quality. It's vital to ensure that the quality of post-decompression data remains high enough for effective processing and analysis. Thus, achieving an equilibrium between compression efficiency and the quality of decompressed data is a critical aspect of seismic data compression.

As depicted in Fig. 6.2, the effectiveness of a compression method can be evaluated using rate-distortion curves, where a curve positioned closer to the lower left corner indicates a more efficient compression algorithm. In the context of current machine learning techniques for seismic data compression, the compression ratio is often pre-determined by setting the size of the embedded vector at the output layer of the encoder. These methods primarily focus on minimizing

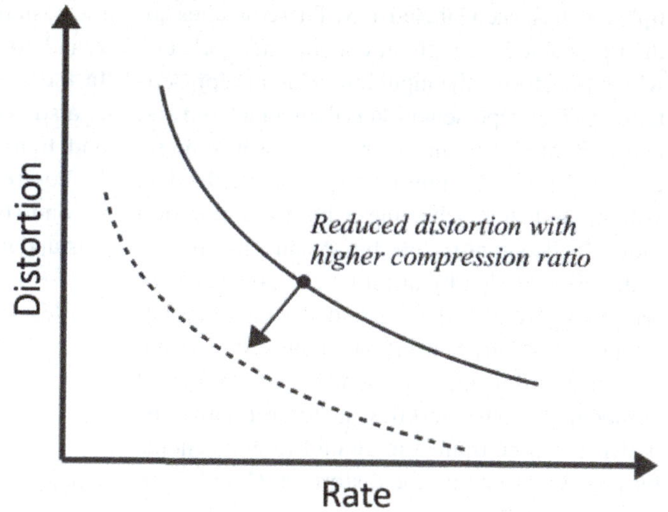

Fig. 6.2 Illustration of the rate-distortion curve. Each curve indicates the balance between the rate and distortion of a specific algorithm varying regarding the configuration

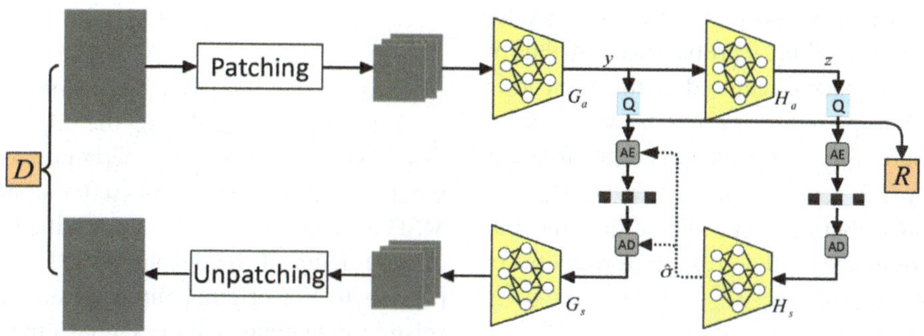

Fig. 6.3 Network architecture. The network is composed of two hierarchy of encoder-decoder pairs, namely $[G_a, G_s]$ and $[H_a, H_s]$. The latent representations from encoders are quantized by **Q**. The arithmetic encoder AE convert the quantized vector into binary streams, which are subsequently decoded by AD. H_s predicts the variance $\hat{\sigma}^2$ for modeling the hidden space of G_s as independent Gaussian distribution. $\mathcal{D}$ represents the distortion between the original data and the decompressed output, while $\mathcal{R}$ computes the entropy loss for estimating the compression ratio

distortion, optimizing only the reconstruction loss incurred by the autoencoder. However, a notable limitation of this approach is the rigidity of the compression ratio, which remains unchangeable once the network's architecture is established. This inflexibility leads to suboptimal compression, especially when dealing with simpler data structures where redundancy is more apparent. To address this issue, a more dynamic approach is required—one that jointly optimizes both the rate (compression efficiency) and distortion (reconstruction fidelity). Such an approach would enable the compression system to automatically adjust, finding an optimal balance between the compression rate and the quality of the reconstructed data, thus catering to the varying characteristics of different seismic data types.

6.2.2 Joint Optimization for the Rate and Distortion with Scale Hyperprior

Building on the advancements in deep learning for natural image compression, we deploy a variational image compression model, as proposed by Ballé et al. (2018) for seismic data compression tasks. The architecture of this network is depicted in Fig. 6.3. The compression process initiates by dividing a seismic gather input into multiple patches, each labeled as $\mathbf{x}$. These patches are then processed through an encoder network G_s, which transforms the input into a latent representation $\mathbf{y}$. This representation is then quantized and encoded into bitstreams using lossless arithmetic encoding methods, significantly reducing the data volume and thus achieving a high compression ratio. To decompress the bit stream, the binary codes are decoded by arithmetic decoding. Subsequently, the decoder G_s and the variance parameter σ^2 jointly convert the latent vector y into the seismic domain, represented as $\hat{\mathbf{x}}$. Since all components in the workflow is differentiable, the network can be trained in an end-to-end scheme by jointly optimizing the distortion $\mathcal{D}$ and compression rate $\mathcal{R}$.

Since the hidden space is losslessly encoded by arithmetic encoding, a probability distribution for the hidden space is required. In this network, a learnable entropy model is constructed to predict the distribution $p_{\hat{\mathbf{y}}}(\hat{\mathbf{y}})$ for quantized latent vector $\hat{\dagger} = Q(\mathbf{y})$. It acts as a surrogate for the real marginal distribution $m(\hat{\mathbf{y}})$. Therefore, the rate loss is formulated as the cross entropy between $p_{\hat{\mathbf{y}}}(\hat{\mathbf{y}})$ and $m(\hat{\mathbf{y}})$ as

$$
\begin{aligned}
\mathcal{R}_{\hat{\mathbf{y}}} &= \mathbb{E}_{\hat{\mathbf{y}}\sim m}[-\log_2 p_{\hat{\mathbf{y}}}(\hat{\mathbf{y}})] \\
&= \mathbb{E}_{\mathbf{x}}[-\log_2 p_{\hat{\mathbf{y}}}(Q(G_a(\mathbf{x})))].
\end{aligned}
\tag{6.1}
$$

The minimal for this loss is 0, meaning the predicted probability is the perfect estimation for the real distribution. And the best compression ratio can be ideally achieved by the entropy-based compression model.

In the most straightforward cases, the distribution $p_{\hat{\mathbf{y}}}(\hat{\mathbf{y}})$ can be a fully factorized format (Ballé et al., 2016). However, to further mitigate spatial dependencies in $\hat{\mathbf{y}}$, Ballé et al. (2018) introduced a hyperprior method to model the distribution of $\hat{\mathbf{y}}$ in a conditional way. Specifically, a prior vector $\mathbf{z}$ is introduced to estimate the variance $\hat{\sigma}^2$) of $\hat{\mathbf{y}}$, which further models the distribution of $\hat{\mathbf{y}}$ into a random Gaussian distribution, i.e. $p_{\hat{\mathbf{y}}|\hat{\mathbf{z}}} \sim \mathcal{N}(0, \hat{\sigma}^2)$. To operationalize this, a secondary encoder, H_a, and decoder H_s are integrated into the system. Their roles are to generate the prior $\mathbf{z}$ and to predict the variance $\hat{\sigma}^2$, respectively. In the compression stage, the quantized prior $\hat{\mathbf{z}}$ is also converted into binary format for storage and transmission. Thus another entropy model is required for estimating the distribution at the second stage. The conditional distortion rate for $\hat{\mathbf{y}}$ is updated as

$$
\begin{aligned}
\mathcal{R}_{\hat{\mathbf{y}}|\hat{\mathbf{z}}} &= \mathbb{E}_{\hat{\mathbf{y}}|\hat{\mathbf{z}}\sim m}[-\log_2 p_{\hat{\mathbf{y}}|\hat{\mathbf{z}}}(\hat{\mathbf{y}}|\hat{\mathbf{z}})] \\
&= \mathbb{E}_{\mathbf{x}}[-\log_2(\mathcal{N}(0, \hat{\sigma}^2) * \mathcal{U}(-\frac{1}{2}, \frac{1}{2}))],
\end{aligned}
\tag{6.2}
$$

where

$$
\hat{\sigma} = H_s(\hat{\mathbf{z}}).
\tag{6.3}
$$

The effectiveness of the reconstruction process is evaluated by calculating the Mean-Squared-Error (MSE) between the uncompressed seismic input $\mathbf{x}$ and the decompressed output $\hat{\mathbf{x}}$. This MSE measurement serves as a critical indicator of the fidelity of the decompressed data in comparison to the original input, ensuring that the seismic data retains its essential characteristics post-compression, i.e.

$$
\begin{aligned}
\mathcal{D} &= \mathbb{E}_{\mathbf{x}}[(\mathbf{x} - \hat{\mathbf{x}})^2] \\
&= \mathbb{E}_{\mathbf{x}}[(\mathbf{x} - G_S(Q(G_a(\mathbf{x}))))^2].
\end{aligned}
\tag{6.4}
$$

In the pursuit of an optimal balance between bit-rate and distortion in seismic data compression, an end-to-end training scheme is implemented by weighted summing the entropy losses and the MSE, facilitating the simultaneous minimization of both terms, expressed as.

$$
\mathcal{L} = \lambda\mathcal{D} + \mathcal{R}_{\hat{\mathbf{y}}|\hat{\mathbf{z}}} + \mathcal{R}_{\hat{\mathbf{z}}}.
\tag{6.5}
$$

In this equation, λ represents the weighting factor assigned to the distortion component. The terms $\mathcal{R}_{\hat{\mathbf{y}}|\hat{\mathbf{z}}}$ and $\mathcal{R}_{\hat{\mathbf{z}}}$ denote the rate costs associated with the quantized representations $\hat{\mathbf{y}}$ and $\hat{\mathbf{z}}$, respectively.

6.3 Numerical Results on Pre-Migration and Post-Migration Datasets

The performance of a compression model can vary considerably when applied to datasets that differ in complexity and noise levels. To comprehensively assess the effectiveness and robustness of our model, we perform tests using both shot gathers and migrated seismic datasets.

For seismic gathers, we use the Mobile AVO Viking Graben line 12 dataset, which has an image size of (120×1500). Each shot gather is gridded and cropped with a patch size of (64×256) for the network. In the case of migrated data case, we use the Parihaka dataset from New Zealand, which has original dimensions of (1226×1168), and we divide the gathers into batches of (512×512). Both datasets are pre-processed using the logarithmic amplitude balancing technique, and are normalized to fit within the range of $[-1, 1]$.

In the optimization of our seismic data compression model, the parameter λ plays a pivotal role in balancing the rate-distortion trade-off during training. This weight is crucial in dictating the model's focus between the accuracy of data reconstruction and the efficiency of the compression ratio. Specifically, a higher λ value emphasizes the preservation of data fidelity in the compression and decompression phases, though it results in a less compact compression ratio. In contrast, a lower λ favors a more efficient compression ratio, but this comes with a trade-off in terms of data accuracy.

Given the unique characteristics of seismic data, such as higher noise levels and imbalanced amplitude distribution, maintaining high fidelity in amplitude preservation is often challenging. To thoroughly evaluate our model's performance in this context, we conducted experiments with five different $\lambda \in \{64, 128, 256, 512, 1024\}$. The decompressed results for our two field datasets are presented in Figs. 6.4 and 6.5. The underlying plots illustrate that, across the spectrum of λ values, the network effectively retains and precisely reconstructs seismic events. Notably, as λ increases, the model places more emphasis on minimizing distortion. This shift results in a reduced residual and, consequently, a higher accuracy in amplitude reconstruction.

The performance variations under different λ settings are quantitatively captured in the rate-distortion curves presented in Fig. 6.6. Each point on these curves corresponds to a model trained with a particular λ value. An increase in λ correlates with a decrease in compression ratio, as evidenced by movement along the bit-per-pixel (BPP) axis. For the pre-migration dataset (Fig. 6.6a), the lowest peak signal-to-noise ratio (PSNR) the model can achieve is approximately 22 dB when the highest compression ratio is obtained at about 38:1, which is calculated by comparing the average BPP of 0.85 after compression against the original size of 32 bits in a 4-byte format segy file. As the compression ratio decreases to 12:1, the PSNR reaches above 29.5 dB. For the Parihaka dataset illustrated in Fig. 6.6b, the model attains the maximum compression ratio of 160:1 with a minimum PSNR of 30 dB. With more weights to the reconstruction fidelity, the PSNR can rise beyond 38 dB at a compression ratio of 73:1. These differences in performance between the datasets highlight the influence of processing stage, noise level, and feature types on the efficiency and quality of seismic data compression and decompression.

6.4 Summary

In this chapter, we delve into a cutting-edge deep learning technique for compressing seismic data. A two-stage variational neural network is constructed with a scale hyperprior to compress the seismic data into binary streams, which are then decompressed to reconstruct the original data with high fidelity. This approach deviates from traditional single-stage encoding-decoding structures, incorporates a hyperprior method for modeling the hidden space distribution, and significantly reduce spatial redundancy in the encoded vector. We assess the effectiveness of this method across different stages of seismic data

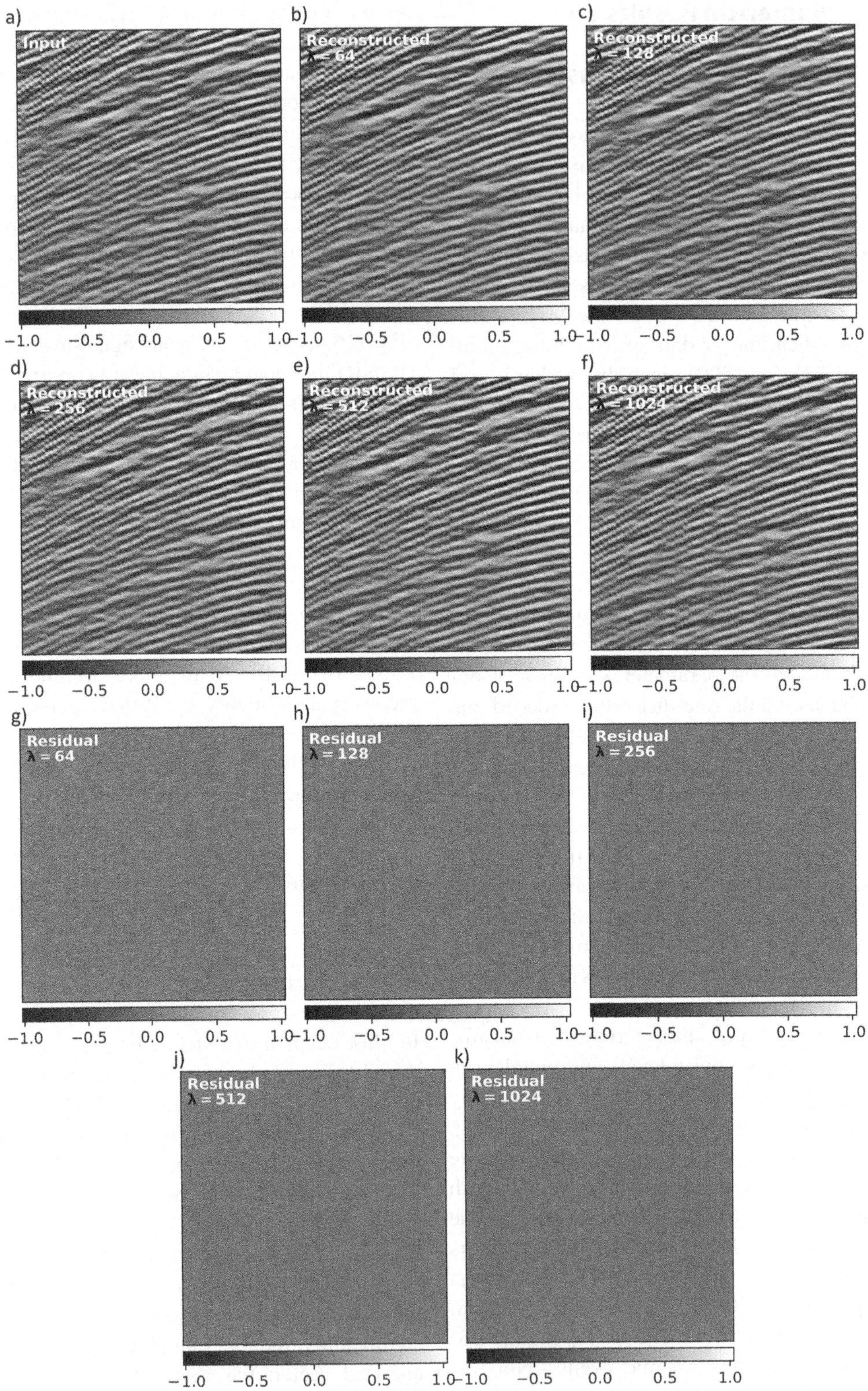

Fig. 6.4 Fidelity comparison of the Mobile AVO Viking Graben line 12 dataset after decompression with different weighting parameter λ. (**a**) The original input; (**b**)–(**f**) Decompression results under λ =64, 128, 256, 512, and 1024. (**g**)–(**k**) difference between (**a**) and (**b**) to (**f**)

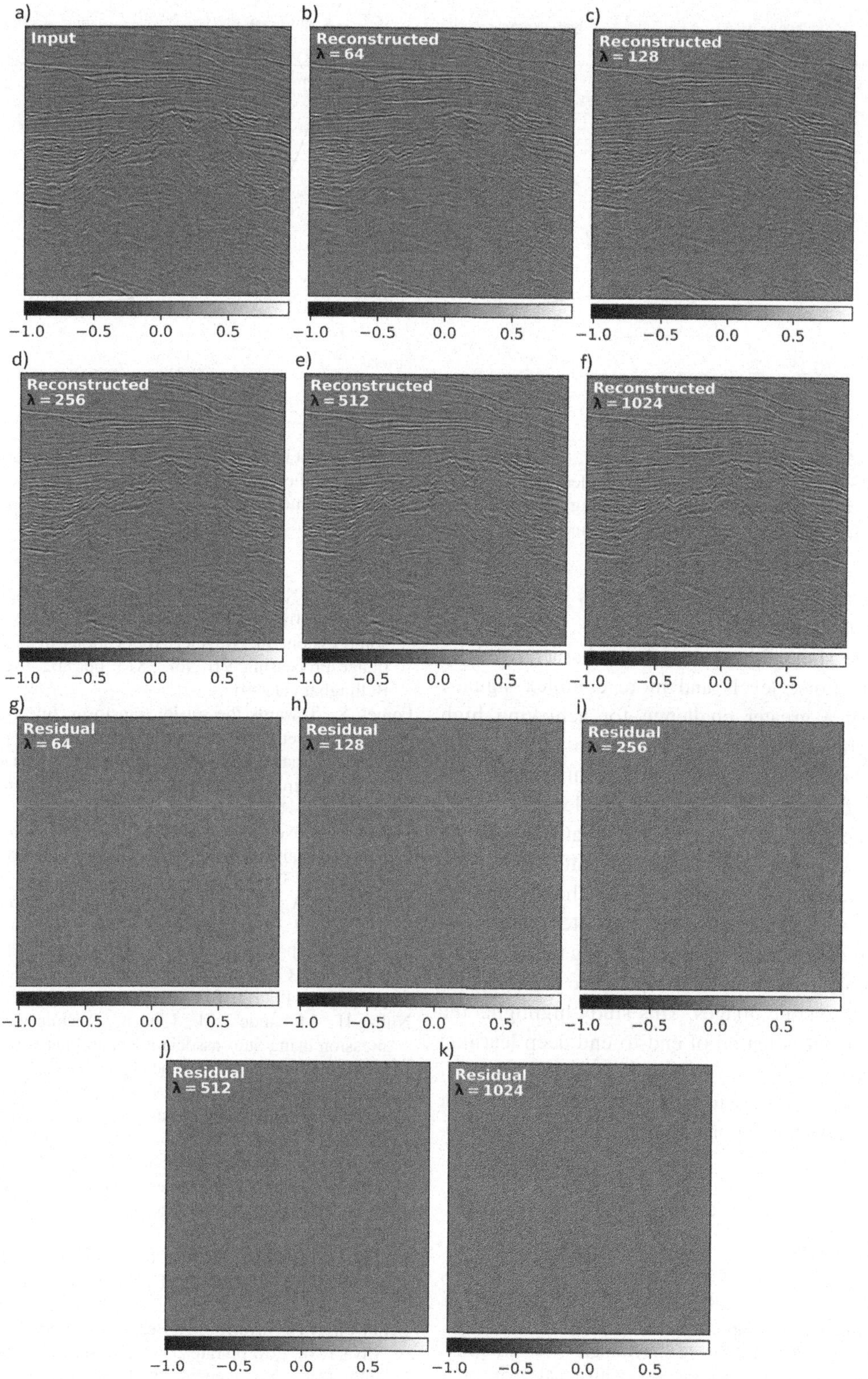

Fig. 6.5 Fidelity comparison of the Parihaka dataset after decompression with different weighting parameter λ. (**a**) The original input; (**b**)–(**f**) Decompression results under λ =64, 128, 256, 512, and 1024. (**g**)–(**k**) difference between (**a**) and (**b**) to (**f**)

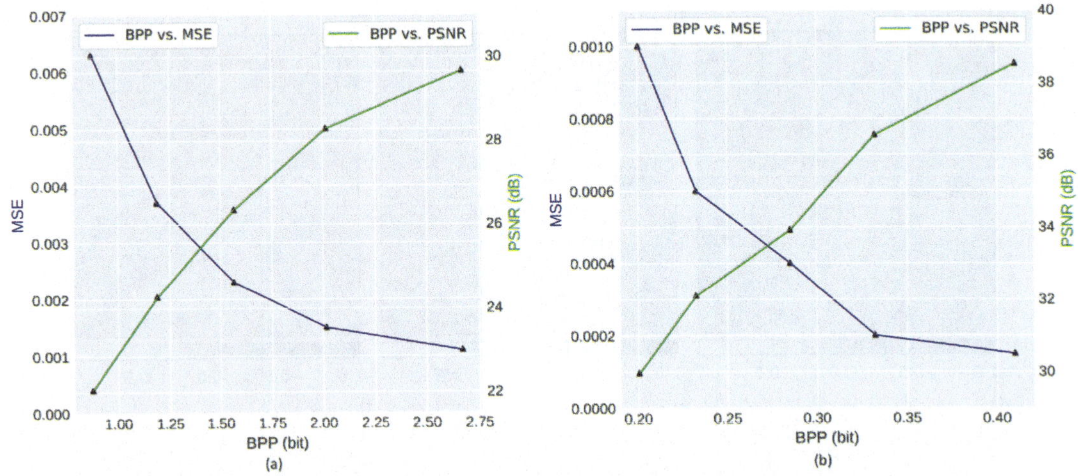

Fig. 6.6 Variation of the R-D trade-off. (**a**) Variation of the MSE (blue) and PSNR (green) under different BPP achievements for the Mobile AVO Viking Graben line 12 dataset; (**b**) Variation of the MSE (blue) and PSNR (green) under different BPP achievements for the Parihaka dataset. Each dot on the curve is obtained from a fully converged compression model with a specified weight parameter λ

processing, including shot gathers and migrated field datasets.

The shot gathers are often characterized by higher noise levels and more complex signals, posing a greater challenge for achieving high compression ratio while maintaining the same level of reconstruction quality compared to post-migration data. Despite these challenges, the experimental results demonstrate that the network can compress seismic data efficiently and effectively at both processing stages. The hyperprior-based variational model demonstrates remarkable compression capabilities, achieving ratios of 11:1 for shot gathers and an impressive 73:1 for migrated seismic images. This study highlights the significant potential of end-to-end deep learning approaches for performing high-efficiency seismic data compression, thus setting the stage for more advanced applications in seismic data processing.

References

Ballé, J., Laparra, V., Simoncelli, E.P.: End-to-end optimized image compression. arXiv:1611.01704 (2016)
Ballé, J., Minnen, D., Singh, S., Hwang, S.J., Johnston, N.: Variational image compression with a scale hyperprior. arXiv:1802.01436 (2018)
Duval, L.C., Nguyen, T.Q.: Seismic data compression: a comparative study between GenLOT and wavelet compression. In: Wavelet Applications in Signal and Image Processing VII, vol. 3813, pp. 802–810. SPIE, Bellingham (1999)
Fomel, S.: Towards the seislet transform. In: 2006 SEG Annual Meeting. OnePetro, Richardson (2006)
Helal, E.B., Saad, O.M., Hafez, A.G., Chen, Y., Dousoky, G.M.: Seismic data compression using deep learning. IEEE Access **9**, 58161–58169 (2021)
Huang, K.Y.: Neural networks for seismic principal components analysis. IEEE Trans. Geosci. Remote Sens. **37**(1), 297–311 (1999)
Mentzer, F., Agustsson, E., Tschannen, M., Timofte, R., Van Gool, L.: Conditional probability models for deep image compression. In: Proceedings of the IEEE Conference on Computer Vision and Pattern Recognition, pp. 4394–4402 (2018)
Nuha, H., Mohandes, M., Liu, B.: Seismic data compression using auto-associative neural network and restricted Boltzmann machine. In: SEG Technical Program Expanded Abstracts 2018, pp. 186–190. Society of Exploration Geophysicists, Houston (2018)
Reiter, E.C., Heller, P.N.: Wavelet transform-based compression of NMO-corrected CDP gathers. In: SEG Technical Program Expanded Abstracts 1994, pp. 731–734. Society of Exploration Geophysicists, Houston (1994)
Rippel, O., Bourdev, L.: Real-time adaptive image compression. In: International Conference on Machine Learning, pp. 2922–2930. PMLR (2017)
Schiavon, A.P., Navarro, J.P., Bernardes Vieira, M., Cruz e Silva, P.M.: Low bit rate 2D seismic image compression with deep autoencoders. In: Computational Science

and Its Applications–ICCSA 2019: 19th International Conference, Saint Petersburg, Russia, July 1–4, 2019, Proceedings, Part I 19, pp. 397–407. Springer, Berlin (2019)

Schiavon, A.P., Ribeiro, K., Navarro, J.P., Vieira, M.B., e Silva, P.M.C.: 3-D poststack seismic data compression with a deep autoencoder. IEEE Geosci. Remote Sens. Lett. **19**, 1–5 (2020)

Spanias, A.S., Jonsson, S.B., Stearns, S.D.: Transform methods for seismic data compression. IEEE Trans. Geosci. Remote Sens. **29**(3), 407–416 (1991)

Wang, Y., Wu, R.S.: Seismic data compression by an adaptive local cosine/sine transform and its effects on migration. Geophys. Prospect. **48**(6), 1009–1031 (2000)

Wood, L.C.: Seismic data compression methods. Geophysics **39**(4), 499–525 (1974)

7 Conclusion

This book has tackled the critical challenge of balancing quality and efficiency in seismic acquisition by developing and implementing innovative deep-learning approaches for the enhancement and representation of seismic data. The proposed methods address various tasks, including low-frequency seismic data extrapolation, seismic data deblending, interpolation, and compression. Given that supervised learning for seismic data processing is often hindered by insufficient labeled training datasets and concerns about the reliability and adaptiveness of the trained models, we have aimed to overcome the model generalization issue by introducing progressive transfer learning and self-supervised learning schema in this work.

For the low-frequency data extrapolation and seismic deblending tasks, we first introduced a novel progressive transfer learning algorithm to iteratively diminish the distributional difference between training and inference phases, consequently addressing the concern for the adaptiveness of deep neural networks. This process relies solely on the raw seismic observations, eliminating the need for prepared labeled datasets from other fields. The experiments on both synthetic and field datasets have proved the effectiveness of the proposed methods.

Seismic data deblending and interpolation can be characterized as under-determined inverse problems. Direct optimization of data misfit using self-supervised learning schemes results in trivial solutions. Thus, we have proposed tailored regularization techniques and constraints specifically for seismic data reconstruction and representation tasks.

The Blind-Trace Network (BTN) has proven to be a highly effective approach for addressing noise leakage issues in seismic data deblending and identity mapping in seismic data interpolation. By explicitly incorporating the coherency of seismic events into the network structure, BTN imposes strict constraints on the inverse problem. Combining the strengths of both filtering-based and inversion-based methods, the hybrid algorithms developed using the BTN structure outperform existing methods in these tasks.

Employing BTN, we have successfully developed a two-stage seismic data deblending algorithm that can manage more complex scenarios with higher blending factors. Our innovative fine-tuning stage, which involves repeating inversion iterations with a converged BTN, significantly enhances the performance.

We have investigated common issues in self-supervised regular seismic data interpolation tasks, including overfitting and aliasing problems. We have proposed two effective approaches, BTN-SS and BTN-mix, both based on the BTN structure, to address these training issues and achieve outstanding results compared to baseline methods. To further address aliasing issues and enhance network robustness and reliability, we have proposed a randomized

S. Wang et al., *Deep Learning for Seismic Data Enhancement and Representation*, Advances in Oil and Gas Exploration & Production, https://doi.org/10.1007/978-3-031-75745-7_7

mixed-training workflow incorporating a dip-informed neural network (DINN) for regular seismic data interpolation. The robust DINN consistently outperforms baseline methods, successfully incorporating dip information into network computation and enabling a seismic-tailored computation scheme.

Lastly, we have examined the application of deep learning for seismic data compression, a critical component of cost-efficiency balancing in data management. Our deep-learning-based method efficiently compresses seismic data at various processing stages, including pre-stack, pre-migration, and post-migration data, with varying compression performance depending on data quality and features.

In conclusion, this book has demonstrated the successful application of novel deep-learning data-processing techniques centering around self-supervised learning for enhancing seismic data quality in the early stages, effectively addressing the cost-efficiency trade-off. We have shown that self-supervised learning for seismic data processing is a promising solution for addressing the data shortage dilemma in subsurface exploration. This approach eliminates the need for laborious labeling and overcomes site-specific limitations inherent in supervised-learning methods. By integrating unique features, mathematical definitions, and physical information in the algorithms, we have successfully developed methods, algorithms, and networks specifically tailored for seismic data representation and processing.

SPRINGER NATURE

GPSR Compliance

The European Union's (EU) General Product Safety Regulation (GPSR) is a set of rules that requires consumer products to be safe and our obligations to ensure this.

If you have any concerns about our products, you can contact us on ProductSafety@springernature.com

In case Publisher is established outside the EU, the EU authorized representative is:

Springer Nature Customer Service Center GmbH
Europaplatz 3
69115 Heidelberg, Germany

The manufacturer's authorised representative in the EU is Springer Nature Customer Service Centre GmbH, Europaplatz 3, 69115 Heidelberg, Germany. If you have any concerns regarding our products, please contact ProductSafety@springernature.com

Printed and bound by CPI Group (UK) Ltd, Croydon, CR0 4YY
30/09/2026
02225259-0020